AF342589

Geoscience Information Society

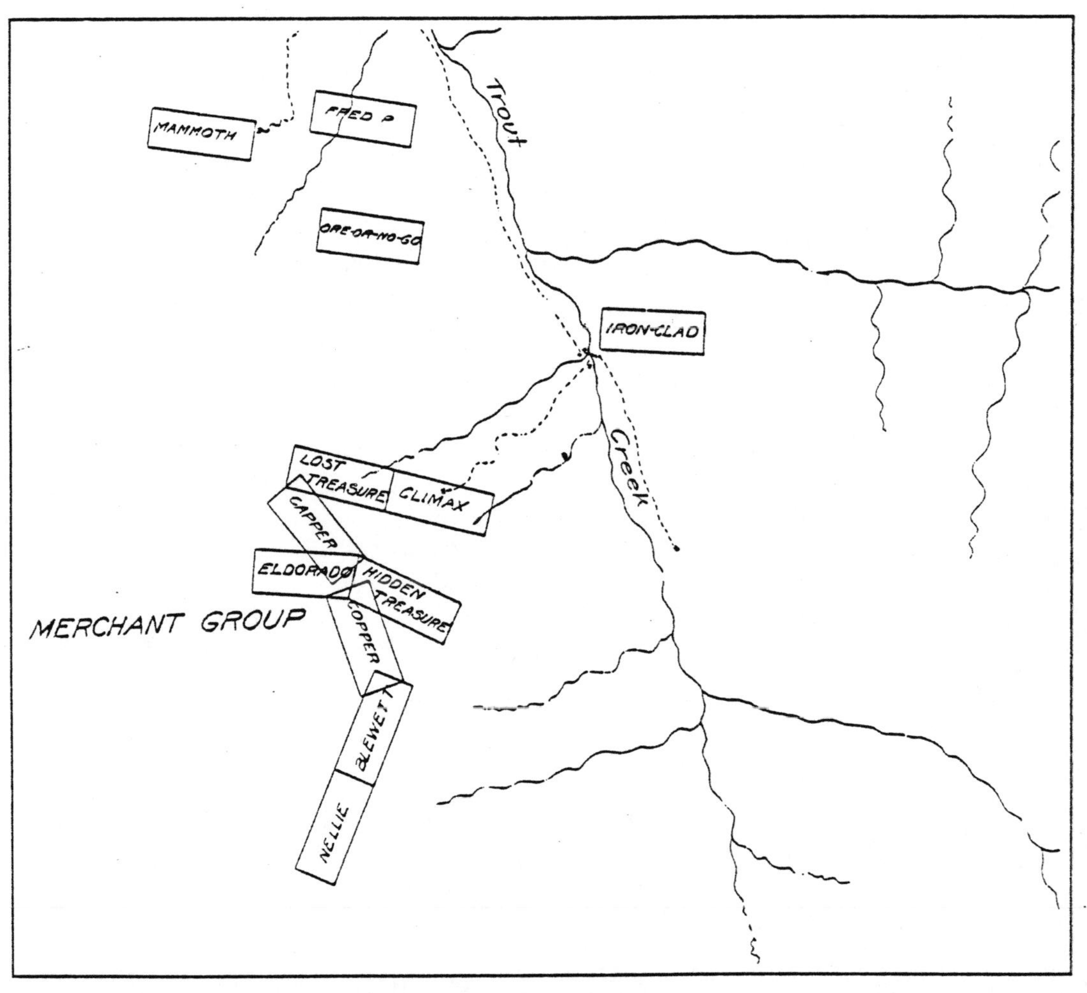

Proceedings
Volume 25
1994

GEOSCIENCE INFORMATION SOCIETY

PROCEEDINGS

25

COVER: Portion of "Map of Mining Claims of the Index
        Mining District Snohomish Co. Wash." Compiled
        by Chas. E. Weaver. IN: Weaver, Charles E.
        (1912), Geology and Ore Deposits of the Index
        Mining District. Olympia, Washington Geological
        Survey. Bulletin no. 7; Plate VII.

PROCEEDINGS OF THE TWENTY-NINTH MEETING
OF THE
GEOSCIENCE INFORMATION SOCIETY

OCTOBER 24-27, 1994
SEATTLE, WASHINGTON

CHANGING GATEWAYS:

THE IMPACT OF TECHNOLOGY ON

GEOSCIENCE INFORMATION EXCHANGE

Edited by
Barbara E. Haner
and
Jim O'Donnell

PROCEEDINGS

VOLUME 25

GEOSCIENCE INFORMATION SOCIETY
1995

For information about copies of this proceedings volume, or earlier issues, contact:

Publications Manager
Geoscience Information Society
c/o American Geological Institute
4220 King Street
Alexandria, Virginia   22032
USA

TABLE OF CONTENTS

PART III

Poster Session

PART IV

GIS Professional Issues Forum

PREFACE

The Geoscience Information Society (GIS) is an independent professional society that was established in 1965. The Society was created to improve the exchange of information in the geosciences by cooperation of an international membership that is now composed of approximately 300 geoscientists and information science professionals. GIS is a member society of the American Geological Institute and an associated member of the Geological Society of America (GSA). The annual meeting of GIS is held concurrently with that of GSA, and papers in geoscience information are presented as a part of the GSA program.

Oral versions of the papers in these Proceedings were presented during the 1994 GSA annual meeting in Seattle, Washington, October 24-27. This volume is divided into four parts:

I. Invited papers, given at the GIS Symposium "Changing Gateways: The Impact of Technology on Geoscience Information Exchange" convened on October 25, 1994;

II. Contributed papers, presented at the GIS Technical Session "Geoscience Information" held on October 24, 1994;

III. Contributed papers, originally on display at the GIS Poster Session on October 26, 1994;

IV. Invited discussion paper, given at the GIS Professional Issues Forum on October 25, 1994.

Papers are arranged within their sessions in the order of original presentation. They have been lightly edited for consistency. The authors are solely responsible for the opinions and ideas.

I thank each of the authors for the time and effort they put into preparing and reviewing their manuscripts. My sincere thanks also to David A. Haner for comments and patience. Finally I thank Lois Heiser, GIS Publications Manager, for handling the printing and distribution of these Proceedings.

This Proceedings volume is dedicated to the late Robert S. Osborne, whose untimely death in July 1994 prevented him from presenting his paper "The Balance Between Technical Proficiency and Scientific Creativity in Postgraduate Education". Bob was particularly keen to be part of the Symposium as he had observed during his many years of teaching at the University of Southern California that students were increasingly computer literate and able to create multidimensional models, but he wondered if they could use and retrieve basic scientific information needed to tackle the complex environmental challenges we face today. Bob prepared for his talk by collecting field data in his own particular style, gathering students together at coffee breaks and lunches for discussions about how they used technology. Around USC and in sedimentological research circles his lively conversations and clear insights will be greatly missed, as he was truly one of those people who acted as a natural gatekeeper and always generated challenging ideas.

Barbara E. Haner
President, GIS
Riverside, California
March 1995

PART I

SYMPOSIUM:

CHANGING GATEWAYS:

THE IMPACT OF TECHNOLOGY ON

GEOSCIENCE INFORMATION EXCHANGE

INTRODUCTION

The 1994 Geoscience Information Society Symposium topic "Changing Gateways: the Impact of Technology on Geoscience Information Exchange" was chosen to reflect the theme of the 1994 Geological Society of America's Annual Meeting in Seattle, geology at "the leading edge," emphasizing dynamic geological processes and showing a scientific discipline at the cutting edge of technology.

Technological changes impact individuals, societies and organizations as they seek to exchange geoscience information and discuss new ideas. Often these changes are enacted through their enthusiastic adoption and utilization by individuals or groups who become experts and gatekeepers, drawing people into new scientific arenas.

It was not until I went to library school that I realized that my own graduate advisor, Donn Gorsline at the University of Southern California, had served me as a major gatekeeper to the world of geology, not only through his inspiring teaching, but also as an active research scientist. He always took the time to introduce his students to colleagues in the network of the invisible college at meetings. He constantly shared new ideas as they emerged through his own work as an editor and a participant in the exchange of preprints and reprints.

Since the mid-1960s, there has been an explosion in information resources with new journals being created, online services, electronic publishing and indexing, and, in the last two years, the potential of providing new communication tools such as electronic bulletin boards and World Wide Web home pages on the Internet. Significant developments in U.S. networking capabilities were enhanced in 1993 when President Clinton outlined the evolution of the National Information Infrastructure to promote long-term economic growth; world leadership in basic science, mathematics and engineering; and the dissemination of federal information in a timely and equitable fashion via a diverse array of sources both public and private, including state and local governments and libraries. Highlighting this movement forward to the new frontier of the information age, magazine covers have displayed headlines such as "Plugged in!" (*U.S. News and World Report*, December 6, 1993); "The Strange New World of the Internet – Battles on the Frontiers of Cyberspace" (*Time*, July 25, 1994); and "Click! Zap! Data till you Drop!" (*Los Angeles Times Magazine*, May 16, 1995).

Papers presented in this symposium illustrate how the technological changes of the last thirty years have altered the way geoscience information specialists, librarians, and geologists have worked, communicated, risen to the new demands of managing data, and exploited the opportunities for incorporating digital technology into the publishing world. These changes have taken place in libraries (Derksen and O'Donnell; Andrews), mapping (Park and Lehmer), management (Hassibe), printing (Aaron and Holoviak), and interpersonal communication (Hallmark). In Britain, there is an old saying, "When one door closes, another opens" encouraging people to look forward positively to new experiences and opportunities. This collection of papers (especially those by John Aaron and Judy Holoviak) clarify the issues that still face the geoscience information community as we adapt to the new challenges of publishing in a digital world where we draw from the printing philosophies of the past and the creative innovations of the brave new world of digital publishing and information dissemination.

Barbara E. Haner
GIS President
Riverside, California
March 1995

WHAT WE DID / WHAT WE DO / WHAT WE'LL DO: GEOSCIENCE INFORMATION CENTERS IN A TIME OF CHANGE, 1970 - 2000

Charlotte R.M. Derksen

Branner Earth Science Library and Map Collections
Stanford University
Stanford, California 94305

Jim O'Donnell

Geology and Planetary Sciences Library
California Institute of Technology
Pasadena, California   91125

Abstract — In the past 25 years, Geoscience Information Specialists have gone from worrying about card filing backlogs to browsing the OPAC; from concern about typographical errors in subject headings to seeing keyword access being our users' preferred method of subject access. In 1970, we worried about shelf space for the *Bibliography and Index of Geology*. In 1975, we knew that, if there was a fire, the one item you saved was the shelflist. In 1980 we were concerned about funding online database searching (and the 300 baud equipment to do it) and the well of state survey gifts was beginning to run dry. In 1985 we were dealing with huge serial cancellations due to inflation, and facing the personal computer revolution and its attendant equipment needs. In 1990 we were still canceling journals and wondering about equipment, this time for developing CD-ROM workstations for GeoRef or loading databases onto local mainframes for unlimited access. In 1994, we're still worrying about space, only now it's digital space.

In an analysis of the subject content of papers presented before the Geoscience Information Society since 1969, we find that trends and topics have remained remarkably similar over 25 years. But in looking at these numbers, we are able to make some generalizations leading to predictions of what lies ahead for the Geoscience Information Center at the millenium.

WHAT WE DID – A CHRONOLOGICAL VIEW

In 1970, when one of the authors was a graduate student in geology at the University of Oregon, using the earth sciences collection was an extremely confusing exercize. Books and journals on the shelf in one area of the library under the call number 550.6 one day were the next week in another section of the library under the call number QE1. This shifting of materials from location to location continued during the whole of her tenure there. The library staff were trying to update and modernize access to the materials and to cope with a continuing space problem. Finding items was extremely difficult, since cards were constantly being pulled from the card catalog for updating.

Using the *Bibliography and Index of Geology* was very tough: while the complex indexing scheme made it imperative to *learn* how to use it, who had time to do that, with tests, field work and lab work? Besides that, items took a long time to get indexed, and once a useful item was identified, interlibrary loan took a long time.

In 1975, as a new librarian in the sciences collection at the University of Wisconsin — Oshkosh, it was a struggle to get people to use the *Bib & Index*. Students found the indexing very difficult to comprehend. Faculty did not want to take time to pore over the paper index; they were clamoring to use GeoRef, which was newly-available online In fact, the geoscience faculty seemed to be pushing for online searches long before those in other disciplines. But online searching was cumbersome and very expensive in those days of 300 baud Texas Instruments print terminals and thermal paper. Each search was crafted as carefully as possible within the confines of the software; even so, retrieval was uneven and generally broader than desired.

Online catalogs were barely in development, so we spent a lot of time teaching students how to search the card catalog and paper indexes. The collection wasn't nearly large enough to meet the teaching and research needs of the faculty. The collection development budget was woefully inadequate. Interlibrary loan was faster than five years before, but still one counted on seven to ten days' wait even for materials from the closest library to come in. Others might take months. Equipment for patron use in the library included a punch card reader as well as some large footprint TI terminals.

By 1980 GeoRef was up on DIALOG as well as on SDC and searching techniques were more sophisticated. Search costs continued to be an issue. There was discussion among librarians as to who should do the database searches: subject experts or search experts — really, the searcher needed to be both. It was still difficult to get adequate search equipment. The chemistry librarian got the first 1200 baud machine, because Chemical Abstracts (CA) was so large and complex that it was important to do those searches on the fastest equipment available. With all of the interest in GeoRef and other databases, there were still many geological survey and society titles (particularly those published in South America, Asia, and Africa) not covered in any indexes. In fact, it is our estimate that 50% of geoscience series titles were not covered by any multinational index at that time.

The first journal cancellation project had already occurred in 1977 - 1978. Still, we remember canceling what seemed like hundreds of journals in 1980 — certainly by then we thought that we had canceled all of the journals that weren't really needed. We were having trouble getting state geological survey publications, as many surveys were no longer sending gift or depository copies to non-state universities. There was trouble trying to find the staff time to place the orders for these materials and so we were trying to streamline order procedures. Two years later, we had assignments to draft storage plans because our shelves were full and there were no plans for new buildings. Even though some places now had online catalogs, many of us were still filing cards, even occasionally typing cards ourselves for field trip guidebooks or IGCP project volumes.

We were searching RLIN and OCLC to facilitate interlibrary loan, still trying to speed up the very slow process of acquiring books and journal articles before the term paper, thesis or grant proposal was due. It was frustrating to finally receive a long-awaited article and realize that the thesis in question had already been completed. We were beginning to turn to document delivery services for some items.

By 1985 we had started a program to teach chemists and geochemists to do their own searches in CA using the academic discount program, and we assisted a few faculty in getting their own Dialog accounts. It always seemed, though, that those who got their own accounts only did one or two searches before returning to the library for mediated searches again. Requests for GeoRef and other database searches were up to over 500 per year. The biggest problem was keeping up with the increasing demand and trying to get the best results possible for the least amount of money. We were trying to get 2400 baud equipment for these searches and downloading the results, which took longer than doing the search, but didn't leave us dependent on the postal service.

For non-core indexes, we were canceling paper subscriptions and planning to rely on online access. Librarians in many institutions were spending considerable time selecting items for remote storage locations. Some were canceling one journal title for every new subscription placed.

Our card catalogs were frozen and we were trying to improve their online replacements. Much of the grant money available to libraries went for projects to convert cataloging records into machine readable form.

By 1990, many of us had small collections of CD-ROMs, and floppy disks of data; we were struggling to find money to purchase the computers on which to run these products. We were also trying to figure out how to run the products and writing handouts for students to explain how to use them. Most of the CD-ROMs had to be searched in the library on stand-alone workstations. Search results were taken away by students in paper form or on floppy disk. We were also trying to convince administrators that we needed to hook these workstations to campus networks so that students and faculty could e-mail or transfer results from the library workstation to their office computers.

We were still trying to find ways to speed up interlibrary loan. We were networking with one another in an effort to collect only part of the geological literature at each institution — to save space, to save staff time, and to save collection money.

Today, most of us have GeoRef and other major indexing tools on CD-ROM, or available via campus online systems, Local Area or Wide Area Networks, or the like. Users, some of them Internet surfers, can tap into the Online Public Access Catalog (OPAC) from offices and homes. We are grappling with what and how much to teach users about Internet resources, and struggling to keep up with software developments and hardware inadequacies, demands and workstation obsolescence.

Collection budgets are still never large enough — journal cancellations occur continuously rather than on a project basis. Space is a premium commodity in almost every library. So far the digital library hasn't produced any noticeable physical space savings; in fact, we now must deal with disk space problems as well.

We are getting more and more of the journal article requests that used to go through interlibrary loan via FAX or from CARL or directly from each other. Many

TABLE 1 . Databases	
CATEGORY	# PAPERS
Building Databases	55
Retrieval of information from databases	38
Description of database/bibliography	11
TOTAL	104

campuses have photocopy services which deliver to the office or the mail box. However, some of this speed is at the cost of heavy staff involvement.

Common themes have haunted geoscience information specialists for years:

- never enough money or staff to acquire all of the material that researchers and students need
- interlibrary loan takes too long
- never enough space to house all the materials that are collected
- a constant struggle to acquire up-to-date, adequate workstations
- incomplete indexes, which can't be easily accessed and used

Technology only seems to heighten our expectations, not to mention those of our patrons. Will this continue to be true into the future?

WHAT WE DID - A TOPICAL VIEW

In order to look at patterns in geoscience information from a less personal standpoint, we reviewed the 327 papers published in the Geoscience Information Society *Proceedings* since volume 1, published in 1969. For this review, we wanted to describe each paper's main thrust with one term, so after consideration, we set aside GeoRef's highly-focused thesaurus and designed our own subject headings.

We developed three main categories to which we assigned most of the papers in the 24 volumes of *Proceedings*: Databases, Traditional Library Topics, and Other.

Databases

104 (32%) of the papers dealt with databases or bibliographies in one aspect or another. The majority of these papers (55) deal with changes in GeoRef, or describe the construction of a specific database (including programming, indexing and collecting data.)

Under Retrieval we put 38 papers dealing with retrieval, either from one database or using a particular system, and papers comparing different databases. We also included papers dealing with how users find databases.

The Description category includes 11 papers on the content of one or more databases or bibliographies.

For each of the first three volumes 40-50 % of the papers dealt with building databases (see Figure 1). This urge to communicate problems, pitfalls and steps in database construction has declined to some degree over the years. There were even four years when there were no papers on database building at all. However, in two volumes more than 50% of the papers were in these areas. A significant number have dealt with the construction of indexes. We have all heard that standardized indexing will become less important, as full text searching and graphical user interfaces become more integrated into the working geologist's world. However, we would counter that continuing expansion

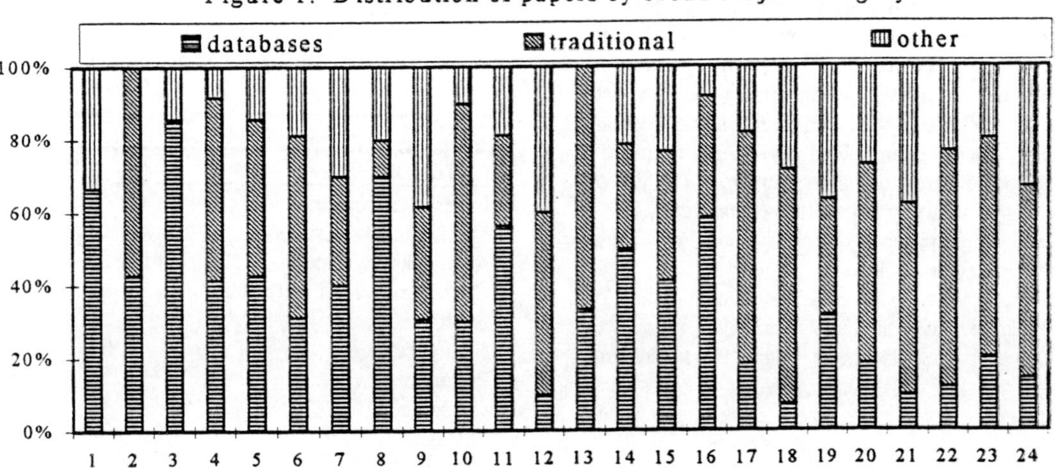

Figure 1. Distribution of papers by broad subject category

of information means that standard indexing procedures within a database will become even more necessary to help deal with mountains of information.

The ability to retrieve needed information from a database has consistently been important. It will continue to be so. The tenor of geoscience research has and will continue to change with technological advances. The practices of indexing will undoubtedly change as both that technology and information science develop. Making databases impart the relevant information has always been of great concern, and conveying to students and faculty Boolean concepts is of prime importance. Research in cognitive processes, machine-aided indexing, user-centered indexing, and graphical user interfaces will all be instrumental in lowering costs of producing databases and at the same time heightening users' retrieval of needed data (Fidel, 1994, Milstead, 1994). Maximizing relevance has been and will continue to be of great interest.

Using databases obviously was and will continue to be a concern, since understanding how the data is accumulated, what data is in which information source, and where to find additional or elusive information is the job of the information professional. However, researchers themselves will continue to be responsible for information on use of sources central to their own work.

TRADITIONAL LIBRARY TOPICS

As expected, the 147 papers included under the heading "traditional" account for the largest group of papers in the *Proceedings*. Surprisingly, there were no such papers published in two of the first three volumes. Papers in this category dealt with the topics listed in Table 2.

Thirty-two papers described information sources, e.g. where to find or how to acquire particular kinds of geoscience information, such as lunar data, geothermal information, and government-published maps. Due to the wide range of literature types, the quantity of materials coming from non-commercial publishers, and the number of foreign published titles, this has been an extremely useful group of papers. As information sources spread across the Internet we expect this line of research to continue for at least the next ten years.

Twenty-six papers described one or more information centers or libraries.

Other librarians talk about the peculiar nature of geoscience literature, and not surprisingly, so do we. Sixteen papers have looked at maps, the nature and quantity of non-English publications, and other such wide-ranging topics. As the number of formats proliferates we can expect to see more discussions on formats, perhaps well logs or laser disks.

A significant number of papers (28) deal with measuring the use of the literature. This is hardly surprising, given the rapidly rising cost of geoscience literature; budgets which have not kept pace with costs; the continuing pressure to keep on our non-expanding shelves only the most-needed materials; and the non-expanding nature of staffing levels in libraries. In fact, the surprise is that there are as few as there are. Hard data on what users say they need to see (in paper or electronically), and how they use what they need, certainly influences which journals libraries acquire, which they get via a document delivery provider, which they put in storage, which they microfilm, and which are digitized. Studies of this type will only become more important as we venture further into the virtual library. We think it will be important for librarians to weigh many factors as the number of avenues to acquire information increases. It will become important to know which serials to continue to get in paper and have on the shelves for real browsing. Other options will surely include subscription to an electronic version which is then mounted on campus, or subscription via the Internet, or access through a vendor such as DIALOG, OCLC, or SilverPlatter.

Given the entrepreneurial, independent geoscientist spirit, some conference proceedings or reference tools will probably be purchased on some medium serving the same function as CD-ROMs or floppy disks do now, or transferred electronically from a remote workstation. There will probably be an increase in the number of titles to which we don't subscribe at all, but for which papers are purchased or acquired on a case by case basis. Given all of these alternatives, use and bibliometric studies will become even more important than they have been in recent years. There will be more factors to weigh, and potentially more dollar savings and less wastage.

Not surprisingly, preservation papers have become more frequent. In 1992, we had a symposium devoted

TABLE 2. Traditional topics	
CATEGORY	# PAPERS
Information sources	32
Specific library descriptions	26
Acquisitions/collection development issues	20
Characteristics of the literature	16
Literature use	14
Bibliometrics	14
Preservation	15
Cooperation	10
TOTAL	147

to this topic. Given the deterioration of older and heavily used materials and the proliferation of new media, this will continue to be a concern for most of us.

Recently, one of the prophets of the rapidly emerging virtual library blithely stated:

> Electronic books mean that libraries need not keep large and expensive stores of bulky and decaying paper ... Libraries will not need to buy multiple copies to allow for book scuffing, book destruction, or to place one book in several categories .... Libraries will not need to chemically treat their decaying books, microfilm them, or transcribe them to Braille, large-print, or audio. All transformations are easier with electronic books" (Rawlins, 1993).

Even if this were completely true, we really doubt that our troubles in this area are over quite yet. There are unreadable tapes of satellite and census data, floppy disks can be erased, and CD-ROMs are not invulnerable. In fact, there is a floppy disk, in a Caltech thesis dated 1989, that cannot be read anywhere on campus. We must continue to do research on, and budget for, preservation of geoscience information for some time to come and include in these plans not only the use of these new media but their preservation as well.

OTHER TOPICS

There are six categories of papers in this section, as listed in Table 3.

Instruction and Education

There has been a lack of rigorous examination of the bibliographic instruction we do. We were surprised at how few papers were about teaching faculty and students to use the library. Most were by faculty discussing how best to teach students various earth science topics, and one discussed how to train library

Table 3. Other topics	
CATEGORY	# PAPERS
Instruction/Education	9
Publishing	21
Earth Sciences & Public Policy	3
Funding/fundraising	3
Information systems/data handling	22
Miscellaneous	18
TOTAL	76

staff. This raises the question of bibliographic instruction in the library. Considering the proliferation of databases available, we can't possibly do in-depth training on all of them. Given the frequent software upgrades and revisions of these products, there seems little point in attempting this. Even if we could offer training, we doubt that there would be many takers. It isn't even possible for most of us to provide explanatory handouts on all of the digital products that the USGS and others provide. Given this situation, what is the value of providing teaching in how to use any database? What if we didn't? What would be lost and by whom? In the digital environment, what should we be teaching students to prepare them for a lifetime of research in the earth sciences?

Publishing

There were no papers on publishing until the society's twelfth meeting. However, since then there have been two symposia devoted to the subject. Given that we are presently in the throes of a revolution in publishing, we expect to see far more discussion in this area. One thing that we feel very safe in predicting is that information is not going to suddenly become cheaper as a result of electronic delivery mechanisms. *Au contraire*, we expect to see the publishers continue to get their money's worth.

This is a time of real opportunity for geoscience societies and university presses: They can regain control of scholarly research output. Physicist Mitchell Golden has recently postulated that "Within a few years, dissemination of research results will go from an expensive process largely controlled by for-profit publishers, to an essentially free, open resource" (Golden, 1994). We disagree: unless some action is taken by societies and universities very soon, publishers will continue to do everything in their power to continue their lucrative practices.

Michael Peskin has made an interesting proposal to the American Institute of Physics. He suggests that an online depository of all physics preprints be established, expanding upon those already set up for some subdisciplines of physics (Peskin, 1994). All papers submitted by society members would become a part of that depository, accessible via the Internet. A certain number of the papers would then be selected for "publishing" in online "journals," giving those the status of refereed papers. It would behoove the various GSA member societies to look carefully at their journal output and rethink their publishing schemes along similar lines that will meet the future needs of their members. Certainly if societies do not step in soon, commercial publishers will be even further ahead and expecting their idea of adequate reimbursement for the products.

Information Systems / Data Handling

Data management and data systems topics include papers on managing a fossil collection, the data manipulation programs used in the deep sea drilling program, use of workstations by geologists, and so on. Due to the importance of large data files to the working geoscientists, we expect to see continued work in this field.

Public Policy, Funding and Fundraising

Neither of these topics was covered nearly as often as we had expected, but we think that will change. In his 1993 GSA Presidential address, Robert Hatcher, said that

> ... our responsibilities as professional geoscientists ... include attaining a level of [the] importance of geoscience with the public; and greater direct involvement in the decision-making processes related to environmental issues, .. resource planning and management (Hatcher, 1994).

We expect more emphasis on public policy in the future and as geoscience information specialists, we should expect to contribute to this effort, particularly those of us in the USGS, state surveys, and other public agencies. Thus we expected a higher number of papers on this topic than the three we've seen. The same is true of papers about fund raising.

Miscellaneous

The miscellaneous category included a variety of topics, as indicated in Table 4.

WHAT WE DO

In the October 7, 1994 issue of *Science*, Robert Pool wrote:

> The Internet has come to resemble an enormous used book store with volumes stacked on shelves and tables and overflowing onto the floor, and a continuous stream of new books being added helter-skelter to the piles.... Almost overnight we've gone from a high-mass, paper-based world to an almost massless, paperless world (Pool, 1994).

Most librarians would agree with the first part of his statement, but few would support the second. True, it seems that we are inundated by e-mail (although much of that announces new resources on the Internet or new tools to help librarians and users to use or at least find those resources). There is very little that is massless or paperless about the average geoscience librarian's workspace.

One impact of new technology is that we can more quickly connect the user with needed information. Another is that the user's knowledge of what is available has grown, although for the most part not as fast as the total which is available, leading to a growing divergence between what is available and what the user knows about. Even so, as the user's knowledge has grown, there is heightened expectation, thereby increasing workload for library staff.

What are we doing now, that we weren't doing in the past? We are (or soon will be) constructing a World Wide Web homepage with numerous links to user-friendly documents that we have prepared, pointers to gophers and other web sources. We're also working on designing and implementing client-server systems; surfing the Internet in search of new resources; learning new systems and preparing point-of-use instruction for a growing array of information sources, and negotiating with vendors for licenses to use information resources (such as flat fees for three passwords for one database, unlimited access to another, buying the tapes and mounting them for another).

TABLE 4. Other Topics: Miscellaneous		
Volume	Year	Topic
7	1977	Foreign language study
11	1981a	Training library staff
12	1981	Third World geoscience info.
14	1983	Trends in geoscience info
15	1985a	History of maps
18	1988	Paleontology collections
18	1988	Rock collections
20	1990a	Field trip guidebooks - 2 papers
20	1990a	History of NEIGC
20	1990a	Evaluate Procite, Endnote, etc.
21	1990	Founding of GIS
21	1990	Meeting the challenge -- statement of where GIS is
22	1992a	Librarians' career expectations
22	1992a	Geoscientists' use of state pubs.
23	1992	Publishing opportunities for librarians
24	1993	Visiting Committee

We face a paradox. On one hand we read in an online announcement that the Library of Congress "plans to announce ... an ambitious effort to convert into digital form the most important materials in its collection and in the collections of all public and research libraries in the country". We read further to discover that LC's "goal [is] to convert the most important materials by the year 2000". In a superb example of understatement, the memo goes on to state that "Copyright is a big issue" (Lewis, 1994). On the other hand we are having a great deal of trouble convincing library administrations to purchase enough hard disk space to load the digital form of some of the map products we've received on depository. Many of us have hundreds of disks and CD-ROMs, but we can't provide any better access than to check them out to users. We will need to work creatively to get from the paper-based library, past the library with digital products and no digital space, to that future library.

WHAT WE'LL DO

Looking into the future is what drew us to this topic in the first place: we never dreamed we'd spend so much time visiting the past. But here we are, making some predictions for the near future.

What will we be doing in the year 2000? Lots of the same things: answering questions, looking for space, developing collections, training and assisting patrons to use tools, getting the needed information to the users, trying to get more, current equipment.

Bibliographic databases will continue to proliferate. Even more digital databases will become available. More and more texts, journals, maps, and books will be available electronically accompanied by pictures, moving images and sound. They will be available via the Internet; via vendors; locally mounted on main frames, on servers, on workstations. Books, serials and journals, however, will still arrive in paper, well past the turn of the century.

Collections will increasingly be in formats other than paper, but we will be "collecting" information sources, finding the places to retrieve needed information, deciding how best to get there, evaluating which of several sources has the most dependable hardware, and which has the most reliable data. In fact we may find ourselves investing in multiple formats because one suits researchers, while another suits the public. For instance we will maintain a collection of topographic maps at the same time as we maintain access to a collection of digital data that can be used to create versions of them. In addition we'll have to a maintain an easy to use Geographic Information System (GIS) so that the occasional user can manipulate this data too.

Cataloging and indexing will become more automated, more complex, and even more important to help users separate the wheat from the chaff.

Archiving and preserving information will be even more vital as we deal with new formats and have more possibilities at our fingertips. It is we information specialists who will be (and should be) held accountable if information becomes inaccessible due to outdated technologies.

While it's logical to expect the librarians of an information-publishing agency to archive that information, that's a new thought for some of us. We have been far more concerned with preserving the purchased items in our collections than we have been with collecting and preserving the reports and manuscripts generated at our own institutions.

Providing access to information resources to users in the field, whether that field is Antarctica, the Himalayas, or the Marianas Trench, should become economically as well as technologically feasible and certainly our responsibility.

Topsy Smalley has said "We have surrendered the common ground shared by print resources for the varietous, shifting ground of online and CD search engines. And we have cast onto the user's shoulders the responsibility for mastering disparate sets of search protocols and commands (Smalley, 1994). To offset this trend there is much talk of developing a seamless information environment, particularly in NSF's Digital Library Projects. However, we see drawbacks. In many ways our environment is already too seamless. Students do not understand why local call numbers are not in every database to which we provide access. A seamless environment suggests that the user doesn't need to know where she is and needn't understand why things are indexed the way they are. But indexing practices vary from source to source; vocabulary varies similarily. The meaning of terms such as porosity, structure, or fault will vary among *Chemical Abstracts*, *Medline*, and *GeoRef*, to say nothing of the *Philosopher's Index*. The fact is that information resources are developed for specific needs and will never be wholly interchangeable.

We need to remember that our purpose is bringing the information and the user together in an efficient manner, not providing great access to lots of databases. We also must be mindful that there are occasional users of any particular tool, and we must make access to that tool relatively painless even for them.

OUR WISH LIST FOR THE YEAR 2000
(OR BEFORE)

Current theses available via server – either from servers

at each university or from a central source such as UMI. There is no technological reason why this can't be done now, since nobody is actually typing theses anymore. Illustrations should be scanned and similarly available.

Technical reports on servers – each agency producing technical reports should be responsible for seeing that they are available on servers.

An easy GIS which can be used by anyone to produce a map in the same amount of time that it now takes to go to a drawer and find the printed map.

Ability to pull documents off and put documents on servers without reformatting, and thus allow the majority of users to pull them off and read them. We must address the compatibility of platforms and formats as well as individual users' varying abilities and capabilities.

Core journals available electronically, at flat fee prices. We think that this will be the next hurdle. It follows a pattern: first we had to pay for database searching, search by search; and we are still working to get flat fee availability for databases. Similarly we seem to be moving towards a world where we pay for journals, article by article. The issue of copyright looms large here.

Photocopiers in libraries (and elsewhere) that automatically scan any document photocopied, for retention into the "library of electronic documents." Again copyright is an issue, although this solution could provide a means of tracking for copyright reimbursement.

Awareness of our users' research and teaching needs continues to be essential, so that we are providing not only what is technologically possible but the desired information in appropriate formats.

CONCLUSION

In 1993, computer scientist Gregory Rawlins wrote:

Soon there will be a whole new profession – people who find things, or know who to ask. Perhaps they will be called ferrets. For those who want to rummage for themselves there will be another new profession – people who arrange things. Perhaps they will be called mapmakers. And everyone will need people who select things – Perhaps they will be called filters (Rawlins, 1993).

Rawlins is right about the kind of activities, but it's not a new profession, it's our profession, and we've been doing these things for decades already. These people aren't called ferrets, or mapmakers, or filters: they're called librarians.

REFERENCES

Fidel, R., 1994, User-centered indexing: Journal of the American Society for Information Science, v. 45, p. 572-576.

Golden, M., 1994, Some thoughts on the future of scholarly publishing: http: // publish.aps.org / archives / eptalk / 0046.html (archived as "Essay on electronic publishing.")

Hatcher, R.D.J., 1994, 1993 Presidential address: GSA Today, v. 4, p. 67-69.

Lewis, P.H., 1994, Library of Congress offering to feed data superhighway: message posted to eechst-l@bgu.edu September 12, 1994.

Milstead, J.L., 1994, Needs for research in indexing: Journal of the American Society for Information Science, v. 45, p. 577-582.

Peskin, M.E., 1994, Reorganization of the APS journals for the era of electronic communication: http:// publish.aps.org / EPRINT / peskin.html

Pool, R., 1994, Turning an info-glut into a library: Science, v. 266, p. 20-22.

Rawlins, G.J E., 1993, Publishing over the next decade: Journal of the American Society for Information Science, v. 44, p. 474-479.

Smalley, T.N., 1994, Computer systems in libraries: have we considered the tradeoffs?: Journal of Academic Librarianship, v. 19, p. 356-361.

APPENDIX 1:
ADDITIONAL READING

Anonymous, 1995, Information revolution calls for changes in behavior: Chemical & Engineering News, v. 73, no.13, p. 22.

Abbott, A., 1992, Modeling the electronic relationship, a scenario with questions: Computers in Libraries, v. 12, no.11, p. 14-15.

Bailey, C.W., Jr., 1995, Network-based electronic publishing of scholarly works: A selective bibliography: Public-Access Computer Systems Review, v. 6, no.1, p. 5-21.

Basch, R., 1991, Books online: visions, plans, and perspectives for electronic text: Online, v. 15, no. 4, p. 13-23.

Bennett, S., 1994, The Copyright challenge: strengthening the public interest in the digital age: Library Journal, v. 119, no. 19, p. 34-37.

Bichteler, J., 1989, Geoscience libraries of the future: predictions for the next decade, in Ansari, M.B., ed., Geoscience Information Society Proceedings, v. 20, p. 97-104.

Broad, W.J., "Doing Science on the Network: A Long Way From Gutenberg," New York Times, May 18, 1993, p. B5, B10.

Cook, K., 1994, The Incredible expanding OPAC: Library Journal, v. 119, no. 21, p. 69.

Dougherty, R.M., and Dougherty, A.P., 1993, The academic library: a time of crisis, change, and opportunity: Journal of Academic Librarianship, v. 18, p. 342-346.

Dougherty, R.M., 1993, Acknowledgment of the past: the first step in changing the future: Journal of Academic Librarianship, v. 19, p. 211.

Drabenstott, K.M., 1993, Analytical review of the library of the future: Washington, D.C., Council on Library Resources, 213 p.

Dunn, L.G., 1993, The Internet, electronic media and changes in geoscience information provision, in Wick, C., ed., Geoscience Information Society Proceedings, v. 24, p. 49-55.

Finkl, C.W., 1983, Megatrends in geoscience information: traditional versus sophisticated technology, in Rowell, U.H., ed., Geoscience Information Society Proceedings, v. 14, p. 109-124.

Fox, E.A., Hix, D., Nowell, L.T., Brueni, D.J., Wake, W.C., and others, 1993, Users, user interfaces, and objects: Envision, a digital library: Journal of Academic Librarianship, v. 44, p. 480-491.

Francica, J.R., 1994, GIS in business for the year 2000 - a preview: GIS World, v. 7, no. 5, p. 52-53.

Garfield, E., 1983, The future of information retrieval in the earth sciences: from bibliographism to encyclopedism, in Rowell, U.H., ed., Geoscience Information Society Proceedings, v. 14, p. 37-53.

Hastings, D.A., 1988, Digital data for interdisciplinary use on workstations: a new opportunity for scientists and information managers, in Lerud, J.V., ed., Geoscience Information Society Proceedings, v. 19, p. 51-56.

Hiller, S.Z., 1993, Geoscience information and libraries: the new paradigm, in Wick, C., ed., Geoscience Information Society Proceedings, v. 24, p. 1-10.

Kerkhof, B.M., 1993, Journal publishing and the (electronic) future: expectations, challenges and reality, in Wick, C., ed., Geoscience Information Society Proceedings, v. 24, p. 11-19.

Kidd, C.M., 1990, Meeting the challenge, in Dvorzak, M., ed., Geoscience Information Society Proceedings, v. 21, p. 195-202.

Lloyd, J.J., 1973, GEO-REF, a report and forecast, in Phinney, H., ed., Geoscience Information Society Proceedings, v. 3, p. 42-44.

Lowry, C.B., 1993, Managing technology perspectives and prospects for a new paradigm: Journal of Academic Librarianship, v. 19, p. 237-238, 246.

McAfee, R., 1990, Early efforts leading to the founding of the Geoscience Information Society, in Dvorzak, M., ed., Geoscience Information Society Proceedings, v. 21, p. 187-193.

McCallum, D., 1994, Brief to the National Information Highway Advisory Council: Newsletter on Serials Pricing Issues, no.123, article 123.1.

Merriam, D.F., 1990, Fax, e-mail, diskettes, softstrip, and CD-ROMs... our communication revolution?, in Dvorzak, M., ed., Geoscience Information Society Proceedings, v. 21, p. 23-32.

Morgan, E.L., 1994, The World-Wide Web and Mosaic: an overview for librarians: Public-access Computer Systems Review, v. 5, no. 6, p. 5-26.

Murray, M.K., 1994, Dissertations go digital as new publishing era dawns: Research Update, no. 26, p. 4-5, 44-45.

Naylor, B., 1994, Willing the future: the twenty percent game: Newsletter on Serials Pricing Issues, no.122, article 122.2.

Perlman, V.A., 1986, In search of the perfect library, in Biers, R.A., ed., Geoscience Information Society Proceedings, v. 17, p. 10-14.

Peters, P.E., 1995, Information age avatars: Library Journal, v. 120, no.5, p. 32-34.

Piggott, S.E.A., 1993, The virtual library: almost there...: Special Libraries, v. 84, p. 206-212.

Reich, V., and Weiser, M., 1993, Libraries are more than Information: Situational Aspects of Electronic Libraries: Palo Alto, CA, Xerox Corporation, CSL-93-21.

Risher, C.A., and Gasaway, L.N., 1994, The great copyright debate: Library Journal, v. 119, no. 15, p. 34-37.

Schatz, B.R., and Hardin, J.B., 1994, NCSA Mosaic and the World Wide Web: global hypermedia protocols for the Internet: Science, v. 265, p. 895-901.

Spilhaus, A.F.J., and Holoviak, J.C., 1990, Society publishing at the millennium, *in* Dvorzak, M., ed., Geoscience Information Society Proceedings, v. 21, p. 55-65.

Stahl, D.G., 1993, The virtual library: prospect and promise or plus ca change, plus c'est la meme chose: Special Libraries, v. 84, p. 202-205.

Stear, J.R., 1974, OASIS, a "one-stop" information service, *in* Morrison, J.L., ed., Geoscience Information Society Proceedings, v. 5, p. 63-70.

Stix, G., 1994, The Speed of write: Scientific American, v. 271, no. 6, p. 106-111.

Veronis, G., 1995, The scientist's role in the publication process: Journal of Marine Research, v. 53, p. i-ii.

Waldrop, M.M., 1994, Culture shock on the networks: Science, v. 265, p. 879-881.

Walker, R.D., 1989, The geoscience journal - its role, past, present and future, *in* Ansari, M.B.,ed., Geoscience Information Society Proceedings, v. 20, p. 87-95.

APPENDIX 2:
DISTRIBUTION OF PAPERS IN GIS PROCEEDINGS VOLUMES 1-24
BY BROAD SUBJECT CATEGORY AND SUBCATEGORY

PROCEEDINGS VOLUME:	1	2	3	4	5	6	7	8	9	10	11	12	13	14	15	16	17	18	19	20	21	22	23	24	Totals
DATABASES																									
Building databases	3	3	3	1	1	4	3	7	1	1	1	1	5	4	1	6		1	4	3				2	55
Retrieval of information from databases	1		2	2	2		1		2	2	7			2	5	1	2		2	1	1	1	3	1	38
Description of database / bibliography			1	2		1			1		1			1	1							1	1	1	11
TRADITIONAL TOPICS																									
Information sources		1		3	1	3					2		7	2	2		1	1		3	1		1	4	32
Specific library descriptions					1	1	1				1	1	1	2		3	4	2	2	3	1	1	1	1	26
Acquisitions / collection development				1					3	6	1	4			1			1		1	1			1	20
Characteristics of the literature		2		1	1										1			2	1	2	3	2		1	16
Literature use				1		2	1								1		1		1		2	2	2	1	14
Bibliometrics						2	1	1					1					2	1	1	1	1		3	14
Preservation													1		2			1	1	1	2		7		15
Cooperation		1							1								1					5	1	1	10
OTHER TOPICS																									
Instruction / education						1		1	1		1			2	1			1						1	9
Publishing									4			2					1	1		1	6	1	3	2	21
Earth sciences & public policy														1		1								1	3
Funding / fundraising														1	1									1	3
Information systems / data handling	2		1	1	1	1	2			1	1	1		1					7	1		1		1	22
Miscellaneous						1	1	1			1	1			1				2	1	4	2	2	1	18
TOTAL PAPERS	6	7	7	12	7	16	10	10	13	10	16	10	15	14	17	12	11	14	19	22	21	17	20	21	327

FROM CARD CATALOG TO COMPUTER FILE:
THE IMPACT OF TECHNOLOGY ON
THE PRODUCTION OF POLAR BIBLIOGRAPHIC DATABASES

Martha Andrews

Institute of Arctic and Alpine Research
University of Colorado
Boulder, Colorado 80309

Abstract — The impact of technology on the production of polar bibliographic databases, which contain significant amounts of geoscience information, is traced from print to Internet. Catalog cards, printed book catalogs, and printed abstracts and indexes, evolved, in the late 1970s and early 1980s, from computer produced printed services into commercial online databases. The personal computer, which arrived in the mid 1980s, allowed widespread access to polar databases online and from 1989 on CD-ROM. The global search capability on CD-ROM files evidenced a significant amount of duplication among separately produced databases, raising concerns about efficient use of resources. It also became apparent that a significant amount of the polar regions literature was escaping effective bibliographic control. Specific instances of overlap and of lack of coverage, were identified on *PolarPac* and *Arctic & Antarctic Regions*, and a plan was formulated to distribute responsibilities for indexing and accessioning the polar regions literature found on these CD-ROMs. The Internet is providing the technology for a current file to be maintained by the database producers, offering real time access by institutional clientele at the same time as the database producers cooperatively build an international bibliographic database with reduced duplication and enhanced coverage.

INTRODUCTION

Polar regions libraries and information services throughout the world have a history of resource sharing directed toward effective service to their specialized clientele. The major resource sharing efforts have involved cooperative acquisitions and interlibrary loan programs, and coordinated development of abstracting and abstracting services. Provision of information by libraries, and by database producers, continues to thrive using the benefits of technology.

The organization of polar information occupies a very specialized niche in the discipline of library and information science. Since at least the 19th century, the polar regions have attracted bibliographic attention as a regional topic, thereby differing from the traditional approach to bibliography through subject discipline or national bibliography.

At present it is estimated that there are some half million online references pertaining to the polar and cold regions. Of these, approximately one third of the research papers and other publications relate to research and development projects involving the geosciences broadly interpreted. The polar regions databases are, therefore, a very significant resource for geoscientists.

Access to the literature of the polar regions has changed dramatically over the past 25 years in response to technological innovations. In the 1960s and 1970s printed card catalogs were transferred through photographic reproduction into published book catalogs. Abstracting and indexing services were making use of computers to print their hard copy. In the late 1970s and 1980s, the bibliographic utilities and online reference databases were mounted on mainframe computers and made available interactively through telephone lines. The advent of the personal computer in the mid 1980s allowed "user friendly" access to electronic databases, both online and on CD-ROM. At present, the disparate storage locations for the polar geological literature are becoming transparent to the users of databases globally searchable via CD-ROM. Meanwhile, telecommunications advances have brought the ultimate networking tool in Internet, which is becoming the vehicle for cooperative database development conducted in a real time, interactive environment.

COLLECTION AND ORGANIZATION OF
POLAR REGIONS INFORMATION
BEFORE ONLINE ACCESS

During the period encompassing the 1960s and early 1970s, the identification, collection, and organization of polar information was controlled by the major libraries and the producers of printed bibliographies dealing with the cold regions literature (Andrews, 1988). Existing libraries expanded, and many new ones were formed (see Andrews and others, 1994 for current list). Catalog cards were exchanged between libraries such as BOREAL and SPRI, and book catalogs, which were basically photographed card catalogs, were printed, published, and shared. G.K. Hall was the publisher of the following catalogs:

Catalogs of the Glaciology Collection, Department of Exploration and Field Research, American Geographical Society, New York (1971)
Research Catalogue of the American Geographical Society, Volume 15, Regional Numbers 47-52. Polar Regions, Oceans, Tropics (American Geographical Society, 1962)
Catalogue of the Library of the Arctic Institute of North America, Montreal, and supplements (1968-1980)
The Library Catalogue of the Scott Polar Research Institute, Cambridge, England and supplement (1976-1981)
Dictionary Catalog of the Stefansson Collection on the Polar Regions in the Dartmouth College Library (1967).

THE DEVELOPMENT OF ABSTRACTING AND INDEXING SERVICES FOR POLAR REGIONS INFORMATION BEFORE ONLINE ACCESS

Following the end of World War II in 1945, strained relations between the USSR and western countries (known as the Cold War) led to increased government spending on research and development with potential use in military operations which were expected to involve the arctic border areas between the USSR and other countries. The demand for production of, and access to, information on the northern circumpolar regions generated thousands of publications and secondary information sources to deal with them (Andrews, 1988). The *Annotated Bibliography on Snow, Ice and Permafrost* (now *Bibliography on Cold Regions Science and Technology*) was first published in 1951; the *Arctic Bibliography*, begun in 1947, was first published in 1953.

For the southern polar regions, the International Geophysical Year (1957-1958) gave impetus to scientific investigation of the antarctic and sub-antarctic regions. A clearinghouse for Antarctic information at the National Science Foundation (NSF), established in 1962, provided the opportunity of organize a coordinated information program for the U.S. Antarctic effort. The first volume of the *Antarctic Bibliography* was published in 1965.

Material for printed indexes was generously shared. Indexers from the larger periodically published indexing services used dozens of libraries in their search for reference materials for:

Arctic Bibliography
Bibliography on Cold Regions Science and Technology
Antarctic Bibliography
Recent Polar and Glaciological Literature
Boreal Northern Titles
Yukon Bibliography
Glaciological Notes.

As indicated above, bibliographical control of the polar literature was a cooperative effort. Access to the literature was broadened through communication and cooperation among individuals and agencies responsible for the accessibility of polar information to an increasingly wide audience.

COMMUNICATIONS BEFORE THE ELECTRONIC ERA

In the days preceding interactive online communication, personal contact at conferences was a very important method of communication. In 1971 a Colloquy on Northern Library Resources [now called the Polar Libraries Colloquy] was held in Edmonton, Canada. Thus began an enduring and still flourishing dialogue among individuals dealing with or having an interest in the collection and dissemination of information on polar and cold regions. The 15th Polar Libraries Colloquy was held in 1994 in Cambridge, England. Between the first colloquy in Canada, and the most recent one in England, this conference was held in the U.S. (both in Alaska and in the lower 48), Finland, France, Norway, and Sweden, with participants from close to twenty countries in both the northern and the southern hemispheres. The colloquy has published a bulletin regularly since its inception, and a third edition of its directory of library resources has recently been published (Andrews and others, 1994).

Other conferences have also played an important part, especially in getting information producers and information users together. In 1974 a *Seminar on the Collection, Transmission and Exchange of Information on the North* was held in Canada. Later conferences sponsored by the Association of Canadian Universities for Northern Studies (ACUNS), and the Northern and Offshore Information Resources (NOIR) Seminar Series, provided valuable interaction. The Alaska (now Arctic) Science Conference, sponsored annually by the

American Association for the Advancesment of Sciences (AAAS), has had frequent information oriented programs.

During the 1960s and early 1970s, most libraries had reasonable expectations of acquiring materials for their collections once they were identified. The numbers of items needed, and their cost, was not yet totally out of reach. Interlibrary loan provided hard copy in its original or xeroxed form to borrowers through the post and telefacsimile. Currently, due to the information explosion and its attendant costs, ACCESS to information is more important than possession at each site.

THE DEVELOPMENT OF THE ONLINE CARD CATALOG

Online Bibliographic Utilities (Technical Services)

A direct descendent of the library card catalog is the online bibliographic utility which started with OCLC in 1971 (Andrews, 1990a). Bibliographic utilities are shared cataloging and technical processing databases. Search only accounts are also available. Public access is limited, with library staff being the main users. The bibliographic utilities provide service for a fee, usually by subscription to the full service, with partial services also available.

All of the bibliographic utilities have some "polar libraries" as members, although not every polar library belongs to a utility.

OCLC (Online Computer Library Center, Inc.) Over 10,000 libraries throughout the world have contributed more than 25 million cataloging records to this utility. The Cold Regions Research and Engineering Library and the Alaska Resources Library are members, along with many academic libraries with significant polar holdings. Foreign language material may be of particular interest also.

RLIN (Research Libraries Information Network). This consortium of 70 research libraries includes Dartmouth College, home of the Stefansson Collection.

WLN (formerly Western Library Network). Among the 500 libraries represented are all of the major Alaskan libraries, and some libraries from western Canada. Therefore, WLN probably has the most bibliographic records pertaining to the polar regions. WLN supports online shared cataloging, and generates several products. This utility allows the Alaskan libraries to distribute cataloging responsibilities among members. The state library, the university libraries, and federal libraries in Alaska all have holdings on WLN.

A recommendation passed at the 12th Northern Libraries Colloquy encouraged polar libraries to become a member of one of the national bibliographic utilities. Full membership in these utilities is too expensive for many of the smaller institutional libraries, but by contributing records to WLN, who then mastered the *PolarPac* CD-ROM, many of the same benefits (cataloging, ILL) become possible.

THE DEVELOPMENT OF ONLINE ACCESS TO POLAR REGIONS ABSTRACTING AND INDEXING SERVICES

The need for polar regions information to be organized on a geographic basis, rather than being sought from discipline oriented databases, has not gone unchallenged. During the 1970s, as printed indexes were transitioning into machine readable indexes, a study was done by the Franklin Institute in Philadelphia which reinforced the need for polar information networks (Franklin Institute, 1977). This study showed a 30% loss in retrieved records if one only looked at discipline oriented databases, most of which are not geographically indexed. BUT geology and geophysics of polar regions were better covered than other polar sciences. It was concluded that the need exists for polar and cold regions information to be provided with a regional rather than a discipline orientation. This need is being met through a variety of services which are being networked to an increasingly large degree.

The influence of online information storage and retrieval technologies on the bibliographic control of polar and cold regions literature has been dramatic. The arrival of online polar/cold regions bibliographical databases was, of course, only a small part of a larger process begun in the 1970s when SDC (ORBIT) and Lockheed (DIALOG) began offering online service via telephone link to mainframe computers remotely located.

The major polar/cold regions abstracting and indexing services were, by the early 1970s, using computers in the production of hardcopy bibliographies (*Arctic Bibliography* and *Bibliography on Cold Regions Science and Technology*). By the late 1970s computer assisted publication of bibliographic references evolved into online access (Andrews, 1990a).

Commercial Online Search Services (Reference)

Most of the "citation" or "article" databases are directly descended from printed indexes and abstracts and are used primarily in reference service rather than in technical services which are the business of the utilities. There are subscription fees and/or charges for online time and type.

The use of several subject-oriented databases is still necessary in polar research.: GeoRef, BIOSIS, NTIS, Compendex, GeoArchive, and ASFA among others.

The beginnings of several of the present online polar/cold regions reference databases were described at the meeting of the seventh Northern Libraries Colloquy, held in Paris in 1978 (Malaurie and Devers, 1982). COLD was, for the first time, publicly demonstrated online and was successfully queried at this colloquy.

COLD. This reference database, compiled at the Library of Congress, is sponsored by the Cold Regions Research and Engineering Laboratory and the National Science Foundation. It is presently online with ORBIT.

BOREAL NORTHERN TITLES, an index of periodicals and newspapers from the Canadian Circumpolar Library, is available online through QL Systems, Ltd.

ASTIS, the Arctic Science and Technology Information System from the Arctic Institute of North America, it is available on QL Systems Ltd.

SPRI, the online catalog of the library of the Scott Polar Research Institute, is available on QL Systems Ltd.

BOREAL, the online catalog of the Canadian Circumpolar Library, is online via CAN/OLE.

Online information services took a quantum leap in the mid-1980s with the advent and wide use of the microcomputer. These services, which had operated from mainframe computers with a batch-oriented retrieval mode, were able to provide online retrieval with all of its attendant features. The microcomputer also allowed in-house databases to be produced, and later shared, with a wider audience.

CD-ROM Technology

CD-ROM technology is completely dependent on the PC (personal computer), with a CD-ROM player attached. Interactive online retrieval with user-friendly software and no per-time-unit charge for use has made this a very attractive technology. The polar regions literature is represented on two specialized CD-ROMs: *Arctic & Antarctic Regions* [NISC (National Information Services Corporation) 1989-] and *PolarPac* (WLN, 1990-).

PolarPac is a direct descendent of a bibliographic utility, in this case WLN. A recommendation passed at the 12th Northern Libraries Colloquy encouraged polar libraries to become a member of one of the national bibliographic utilities. By contributing records to WLN, who then mastered the *PolarPac* CD-ROM, many of the same benefits (cataloging, ILL) become possible. WLN already had, in 1987, introduced a large portion of its database on CD-ROM as LaserCat. Using the same software, and selecting records contributed to WLN by many polar libraries, in 1990 *PolarPac* was first published and has been updated twice already. Development funds have been granted by NSF, NEH, and USIA.

Arctic & Antarctic Regions (*AAR*) is not so much a descendent of the online reference databases as it is a new and better medium for their presentation and access. Five of the seven reference databases mounted on the September 1993 edition of *AAR* were already available online (the other two were developed as in-house databases). However, the five databases were mounted by three different vendors using separate software, while CD-ROM with its globally searchable environment provides very convenient and user friendly access.

The global search capabilities on the CD-ROMs allow a level of comparability between databases which was very difficult to achieve in earlier studies. These capabilities made it possible to engage in the study described next. With global search technology, duplication of effort becomes evident. Users are uncomfortable with this since it is not always obvious to them that duplicates are what they seem, and funding agencies are very uncomfortable with the appearance of wasted effort. Analysis of the CD-ROMs, described below, has resulted in identification of problem areas and recommendations for redress.

INVESTIGATION OF JOURNAL ARTICLE OVERLAP (AND COVERAGE GAPS) IN POLAR AND COLD REGIONS DATABASES

The issue of bibliographic control of the polar regions literature has been discussed and written about in association with the Polar Libraries Colloquy (PLC) for over twenty years. In fact, some of the material presented here has already been accepted for publication in the most recent PLC proceedings (Andrews, 1994, in press). Four years ago at the 13th Polar Libraries Colloquy I presented a conceptual diagram illustrating the notion that the abstracting and indexing services are providing duplicate coverage of a portion of the polar regions literature which is approximately equal to the amount of the literature that is not being covered at all (Andrews, 1990b). This concept has subsequently been shown to be fairly accurate.

This was only a concept, an estimate, since before the advent of globally searchable databases on CD-

ROM, not only was it difficult to assess the situation as regards overlap and coverage gap at an article level, but it would have been difficult to persuade database producers to change selection policies which were designed mainly to accommodate their institutional clientele.

Recently I have been able, through a grant from the Council on Library Resources (Andrews, 1993), to quantify the incidence of overlapping coverage, and of gaps in coverage, of five databases provided on *AAR*. This grant also provided for a related study of *PolarPac* which resulted in some recommendations for shared collection development.

The goals of this project were twofold: to reduce the overlapping bibliographic coverage of polar regions literature by indexing services, and to expand coverage to include document types not currently selected by those services.

Coverage by the databases included on these CD-ROMs is remarkably comprehensive, and access is user friendly. However, there is a high incidence of overlap in the materials selected by the various producers whose databases are globally searchable on *AAR*. At the same time, there is a gap in coverage for a high percentage of published information which could reasonably be expected, because of its polar regions orientation, to be indexed on one or the other of the two CD-ROMs.

The systematic research undertaken for the Council on Library Resources project, "Distributing responsibilities for accessioning and indexing polar regions information," (Andrews, 1993) resulted in the specific identification of problem areas associated with both overlapping coverage, and lack of coverage, of the polar regions literature. Prior to this research, these problems were only generally known, and the several studies done were difficult to compare with each other (Andrews, 1993, p. 13-15). Secondary (abstracting and indexing) services have evolved independently of each other; these services were targeted to a specific user community and there was no incentive for these services to coordinate their selection policies early on.

Both sides of this problem are worthy of redress in order for information providers to serve their user community responsibly. Public and private resources are being diluted by duplicate efforts, while the user must seek additional access sources which are not easily identifiable and/or available. Improved service to the users of the polar regions databases should be an important goal of polar information providers.

To achieve the project goals an analysis was undertaken of coverage included in the March 1991 and May 1992 editions of *AAR* and of Issue 2 (1991) of *PolarPac*, using statistical data derived from the discs, and statistical data supplied by WLN.

Overlapping coverage of journal articles

Overlapping coverage between databases published on the March 1991 edition of *AAR* has already been established to some extent by the use of compositing software by the publisher. To identify this duplication more specifically, I established a routine using INMAGIC and Excel software. Eighteen journals were selected and tested for coverage by all five databases (COLD, SPRI, ASTIS, C-CORE, and WDC-A) globally searchable on *AAR*. Overall, of articles published in 1988 by these eighteen journals, 331 articles were selected and covered 638 times (1.92 times each) by the five databases. COLD and SPRI had both selected 145 of the same articles. Due to this high incidence of duplication, COLD and SPRI were chosen for further comparison.

What emerges is that some journals are covered extremely well and by more than one service. The producers of both COLD and SPRI are now, as they have been in the past, aware of their duplication of effort. With the availability of COLD on Internet (see below), allowing real time access, both producers now have the mechanism to cooperate and distribute responsibilities for coverage. In addition, the producers have enough empirical evidence to support changes being made.

Gaps in coverage

The polar regions CD-ROMs were next analyzed to determine their effectiveness for retrieval of a wide range of subjects that could reasonably be expected to be found since they all represented scientific research conducted in the arctic regions. Resolution of this issue is probably even more important than the issue of duplication. A checklist of 630 titles was compiled from reference lists of fifteen review papers on polar regions topics. Of these 630 titles, 364 were found, and 266 were not found on *AAR*, May 1992 edition. When only these 266 titles were searched on *PolarPac*, 49 were found. Further analysis of the 266 titles not found on *AAR* shows that approximately 20% each of technical reports, theses/maps/abstracts, book chapters/conference papers, and nonpolar journal articles are not indexed.

Of the 49 records (of the missing 266) subsequently found on *PolarPac*, 32 were technical reports (half of the 64 "missing" on *AAR*), and the other 17 were scattered almost equally among the other categories.

What emerges is that unlike the well covered peer-reviewed journals, other journals, and non-journal publication types, are poorly covered.

The 364 found items (630 minus 266) were indexed a total of 746 times (or 2.0 times each). This number is comparable to the 331 articles (overlap, above) having been indexed 638 times (or 1.92 times each). It gives weight to the notion that if resources were reallocated, the effort consumed in duplication could be used to expand coverage to a more satisfactory level.

Study Conclusions

Selection and indexing policies assume a different level of importance in a globally searchable format. The polar regions CD-ROMs are being used by an international community whose needs must be met. Improved bibliographic control of polar regions information is vital because this information forms the basis for policy and management decisions, further polar regions research, and planning data for residents of the northern circumpolar region.

The results of the CLR project demonstrate that the peer-reviewed journals publishing mainly polar regions material are very well covered by the major reference databases; in fact there is considerable overlap in the coverage. On the other hand, discipline oriented journals, and non-journal document types, are not covered well at all on the CDs, leaving the user to access valuable research materials through other channels.

THE FUTURE

Networking and database production

The producers of the two major polar regions databases, COLD and SPRI, have agreed to jointly develop a current, working file on the Internet, through which they will be able to reduce the amount of duplicated effort. Each producer will be the sole source indexer for specified journals, and the resulting file will be attached to both of their databases. Both databases will still appear in a globally searchable file on *AAR*, and the user will only retrieve each record once; the CD-ROMs producer will have already identified these common records as composites.

When this project, now underway, is running smoothly, these two database producers may be able to allow further remote input, with good quality control, which will expand coverage to include more theses, abstracts, nonpolar journals, and technical reports. Remote input on the Internet will be the technology which will allow true distribution of indexing responsibilities among polar information providers.

Communications and information access

The Internet is the ultimate facilitator for networking. Library networks grew out of sharing activities, such as cataloging, interlibrary loan, preparation of indexes, and even cooperative acquisition and storage of information. The Internet now allows library catalogs and indexes to be browsed remotely.

In late 1993 a LISTSERV Distribution List POLLIB-L was initiated for those involved with polar libraries and polar bibliographic databases. This LISTSERV resides at the University of Calgary and is managed by Eric Tull and Ross Goodwin. Recently Eric Tull began developing a Polar Information Sources Gopher which is growing rapidly with input from many POLLIB-L members.

This gopher contains directories of arctic data and information resources, and libraries. It also provides access to some library catalogs. Its future potential is limitless, and will not doubt be heavily used in the future.

CONCLUSIONS

Computer networks are playing a major role in the research community. The fact that many elements for a polar information network, as outline in this paper, exist and are linked in various degrees already, provides realistic expectations that the information needs of the polar and cold region information user community will increasingly be met through geographically indexed bibliographic services available through telecommunications networks and CD-ROM.

In the foreseeable future, polar information delivery will continue to rely heavily on electronic networking of regionally produced components, i.e., library catalogues, online reference services, bibliographic utilities, and CD-ROM products. The role of print technology will not disappear as long as several older but still valuable reference works exist only in print form. Commercial online access remains the best provider of current information, and some online databases are not available in other forms. CD-ROM, because of its reasonable subscription rate with no online charges, has many advantages for information provision to smaller libraries and individuals. The Internet will probably grow in importance, but "navigating the Internet" still poses many challenges.

Technological change is inevitable and the polar information community will be on the cutting edge of implementing change to benefit production of, and access to, databases for the polar regions scientific community.

bibliographic databases came from the Council on Library Resources Grant 897. The National Science Foundation (NSF) Grant DPP DPP8913041 supported research on the use of CD-ROM technology for a U.S. Polar Bibliographic Information System. NSF is currently supporting the U.S. Polar Information Working Group in their program to serve the information needs of the polar community through NSF Grant OPP-9321320.

REFERENCES

Andrews, M., 1988, The organization of polar information before the advent of online databases: a review of the literature: Special Libraries Association Geography and Map Division Bulletin, no.154, p. 5-22.

Andrews, M., 1990a, Computerized information retrieval and bibliographic control of the polar and/or cold regions literature: a review: Special Libraries Association Geography and Map Division Bulletin no.159, p. 21-42.

Andrews, M., 1990b, Resources sharing among U.S. providers of polar information, *in*: Kivilahti, R., Kurppa, L., and Pretes, M., eds., Man's future in Arctic Areas: Proceedings of the 13th Polar Libraries Colloquy, 10-14 June 1990. Rovaniemi, Finland, University of Lapland, Arctic Centre Publications, no. 1, p. 48-51.

Andrews, M., 1993. Distributing responsibilities for accessioning and indexing polar regions information. Final Report to the Council on Library Resources, Grant 897. 5 p. + 6 Appendices.

Andrews, M., 1994. Are we information poor? Limitations to accessing polar literature, *in press*, Proceedings of the 15th Polar Libraries Colloquy.

Andrews, M., Brennan, A., and Kurppa, L., 1994, Polar and cold regions library resources, a directory. 3d. edition: Boulder, Colorado, Polar Libraries Colloquy, 208 p.

Franklin Institute Research Laboratories, 1977, Study of coverage of arctic research by discipline-oriented abstracting and indexing services. Final Report (to NSF). Various pagings.

Malaurie, J. and Devers, S., eds., 1982, Arctica 1978. VIIe Congres International des Bibliotheques Nordiques. 7th Northern Libraries Colloquy. 19-23 spetembre 1978: Paris, Editions du Centre National de la Recherche Scientifique, 581 p.

National Information Services Corporation, Arctic & Antarctic Regions [CD-ROM], 1989-   .

WLN (formerly Western Library Network), PolarPac [CD-ROM], 1990-   .

CONSTRUCTION AND DISSEMINATION OF A DIGITAL FAULT DATA BASE FOR THE SOUTHERN CALIFORNIA EARTHQUAKE CENTER

Stephen K. Park
Eric Lehmer

Geographic Information Systems Laboratory
Southern California Earthquake Center
University of California
Riverside, California 92521

Abstract – A key task of the Southern California Earthquake Center (SCEC) is the development of a four-dimensional (three spatial coordinates and time) model of the earthquake generation process in southern California. Such a model will help better assess the effects of future earthquakes and identify new hazards. A geographic information system (GIS) provides the framework for the construction of this Master Model and allows researchers to search for previously unrecognized relationships between diverse types of data. Incorporation of information such as geologic maps, faults, geotechnical data, earthquake parameters, and topography have provided new challenges in database design and construction. For example, many sources of faults are available and the appropriate choice depends the potential use of the database. Scale, fault activity, and types of fault attributes depend on the research undertaken. It is generally not possible to design a database for all users, and the most successful applications have occurred when the researchers have been involved in all phases of database design and construction. However, the availability of these data in digital form permits misuse through lack of appreciation of the limits of such data. Additionally, dissemination of earthquake databases requires a review process much like peer review of journal articles but such procedures are still under development by SCEC.

INTRODUCTION

The use of geographic information systems (GIS) has permitted researchers to examine spatial relationships between disparate data sets, allowing for more sophisticated pattern recognition. While such pattern recognition has been the basis of geological and geophysical research for years, the conversion of analog data (i.e. maps) to digital form greatly expands the possibilities for correlations. The tedious tasks of representing many different data on common scales with common projections can now be done with a few key strokes at a computer terminal. However, there is a hidden cost for this expanded research capability. Data now in analog form must be converted to digital data, and this process is time intensive. This cost is appreciable, and requires choices regarding what data to include, what scale for compilation, and how to distribute the results.

We will illustrate the advantages, and pitfalls, of the use of GIS in geoscience research with an example of a fault data base created for the Southern California Earthquake Center. The advantages include integration of different data sets, rapid calculation of derivative products, distribution of results in readily usable form, and ease of updates. Disadvantages include questions of copyright of material, lack of forum for publication of digital results, and responding to disparate needs of users.

The Southern California Earthquake Center (SCEC) is a consortium of academic, government, and commercial researchers whose task is to develop a four-dimensional (three spatial dimensions and time) model of the earthquake process in southern California with the ultimate goal of reducing earthquake risk through better estimation of seismic hazards. SCEC is a center without walls, and scientists are linked through Internet and monthly seiminars. Therefore, exchange of geoscience data in digital form is essential to the collaboration of the member scientists. GIS provide a ready format for comparison and exchange of those data.

PROBABILISTIC SEISMIC HAZARD MAPS

Construction of a fault data base requires that several decisions be made at the outset. First, a compilation scale must be chosen. Second, which faults should be digitized? Third, what information other than geographic location will be included? All of these questions are dependent on the most important question: How will this fault data base be used? We answered this question by referring back to the mission of SCEC to better estimate seismic hazard in southern California. This mission was the primary use of the fault data base. As an example of how this task might be accomplished, we developed a program for estimating probabilistic seismic hazard patterned after the method outlined in Wesnousky (1986). This method predicts the probability that horizontal ground acceleration at a point will exceed a certain level in a set time period. A probabilistic seismic hazard map differs from an earthquake scenario because it considers the effects of multiple faults and assigns a probability to the ground shaking.

Fault location

Given the uncertainties in estimation of probabilistic ground shaking, very general fault maps would probably suffice for this task. SCEC's task is to improve and refine our hazard estimation however, and choice of compilation scale based on our current knowledge would be inadequate. Additionally, geologists using this data base for other research require a fault data base at as detailed scale as possible. We therefore chose to compile the fault data base at the most detailed scale possible.

This decision must be tempered by the availability of data, however. Analog geologic maps at scales of 1:750,000 and 1:250,000 are available (Table 1) but are at too coarse a scale. Many faults on these maps are no longer active and contribute nothing to the seismic hazard. Alquist-Priolo (A-P) ground rupture maps are available for the state at a scale of 1:24,000, but map only those faults that have ruptured the ground in the last 10,000 years (Table 1). Studies of individual faults at more detailed scales than 1:24,000 (i.e., 1:12,000) are available, but do not cover southern California uniformly (Table 1). Uniform coverage at a given scale is crucial because digital data bases can be used to generate maps at any scale. Generalization of a finer scale to coarser one will not result in significant errors of location, but the reverse is not true. Extrapolation to a finer scale can result in appreciable errors of location. Mixing of data from different scales results in a data base that is only as good as the coarsest scale. Faults are only incompletely available at scales finer than

1:24,000, and supplementing these data with the A-P faults would result in a data base that is strictly only accurate at scales of 1:24,000 and coarser. Therefore, we elected to base our digital faults on the A-P maps because they represented the most detailed uniform coverage for southern California. These rupture zones were sampled with an approximate digitization interval of 100 m.

Most of the faults identified by Wesnousky (1986) as contributing to seismic hazard are shown on these maps. Active faults from Wesnousky (1986) not on the A-P maps were digitized from the state 1:250,000 geologic maps (Table 1), and constitute approximately 25% of the data base. While this mixing of scales does degrade the spatial accuracy of the data base, the reduction in accuracy is probably not significant because of the small percentage of faults from the 1:250,000 maps and because they contribute only marginally to the hazard.

Fault segmentation and maximum magnitude

The analog A-P maps digitized for the fault data base consist of many short rupture segments (Figure 1). Many of these ruptures resulted from a single earthquake on a single fault and must be grouped accordingly for the following reasons. Observations show that peak ground acceleration is dependent on the distance from the earthquake source and to the magnitude of the earthquake (Joyner and Fumal, 1985). Additionally, there is a positive correlation between rupture length and magnitude (i.e., Aki and Richards, 1980). Therefore, single ruptures from the A-P maps must be grouped into fault segments (Figure 2) in order to provide accurate estimates of maximum magnitude and to determine the distance between the observation points and the faults.

Fault segmentation is a controversial issue. Based on geologic mapping, observations of historic ground rupture, and analysis of instrumental and historic earthquake data, different researchers have segmented the same faults differently (i.e., Wesnousky, 1986; Clarke and others, 1984; Ward, 1994). Some geologists argue that even assigning fault segments is incorrect. In the compilation of the fault data base, our choices were to do our own segmentation (adding yet another analysis

TABLE 1. Available analog fault maps

MAP	SCALE	FAULT TYPES	SOURCE
Fault map of California	1:750,000	All faults	Jennings (1977)
State geologic maps	1:250,000	All faults	Rogers (1965)
Alquist-Priolo maps	1:24,000	Holocene rupture zones	Hart (1992)
Individual studies	various	various	theses, papers, etc.

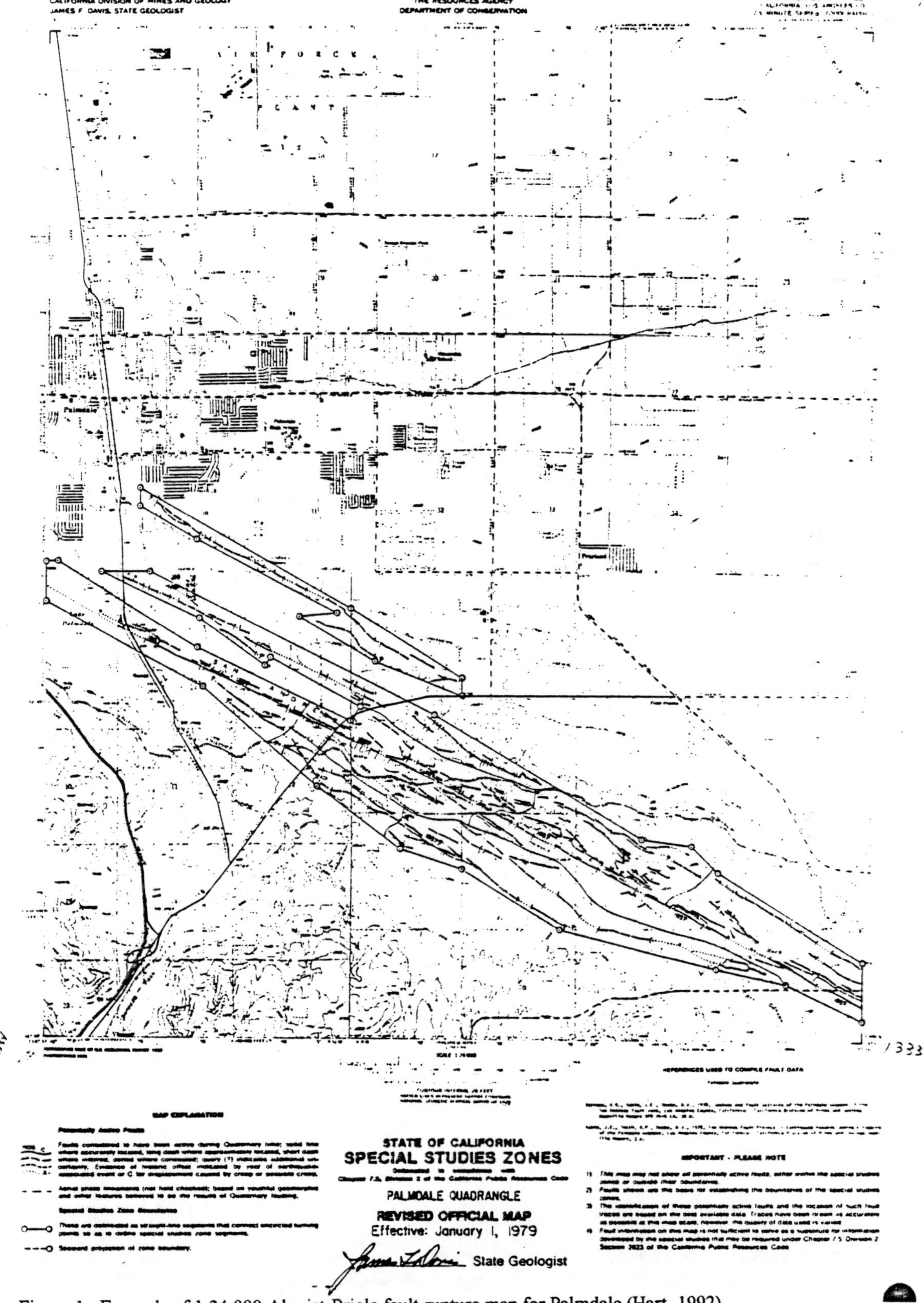

Figure 1. Example of 1:24,000 Alquist-Priolo fault rupture map for Palmdale (Hart, 1992).

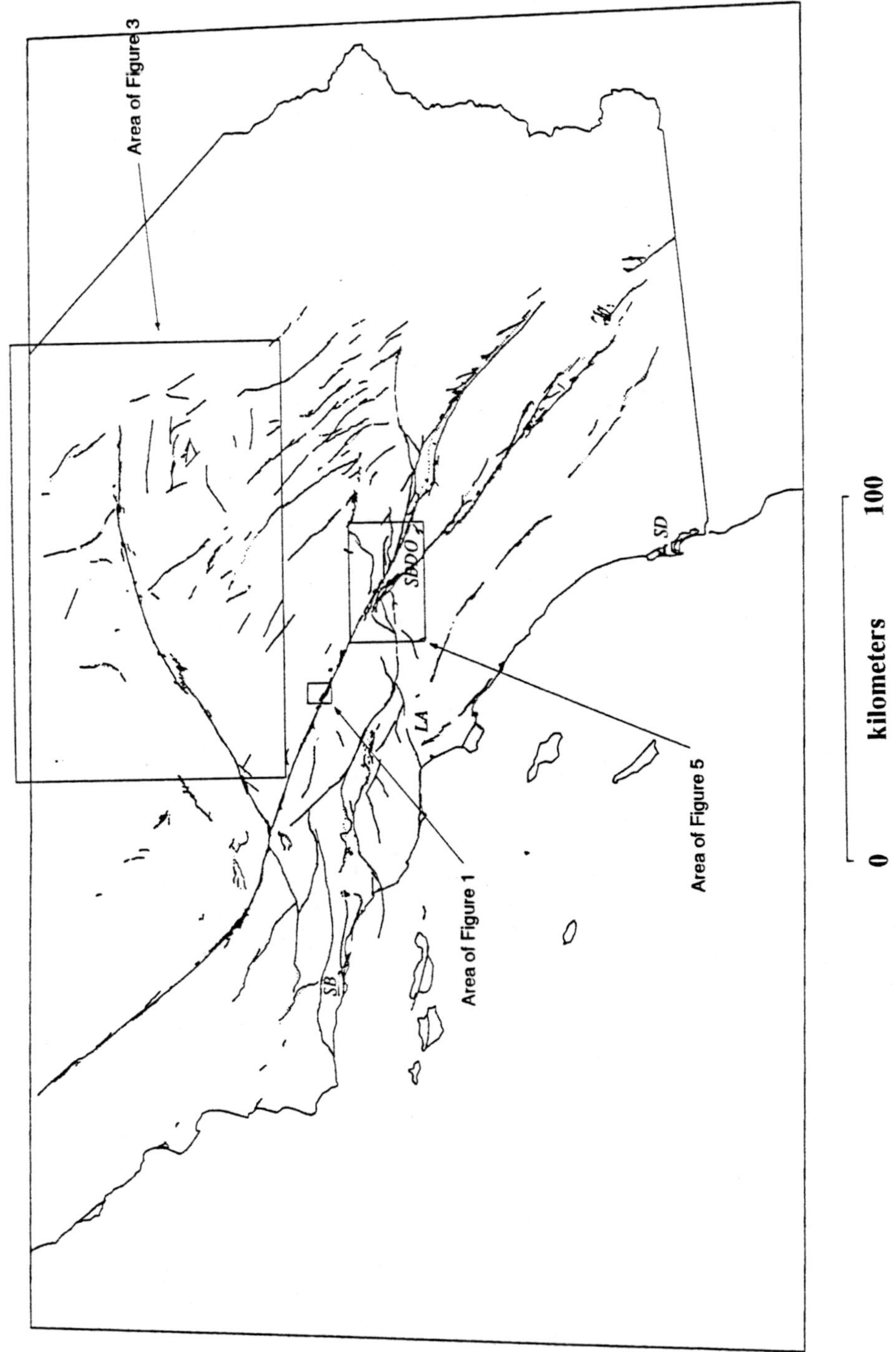

Figure 2. Fault rupture zones and generalized faults created from grouping ruptures into fault segments. Areas of Figures 1, 3, and 5 are shown. Symbols used are: Los Angeles, LA; Santa Barbara, SB; San Bernardino, SBDO; and San Diego, SD.

to the collection) or use one of the published ones. Use of a published one has the advantage that it has been peer-reviewed, and we chose to use Wesnousky's segmentation (1986). An important feature of the fault data base is that this segmentation can be revised easily by simply combining the A-P ruptures into different groupings.

The empirical relationship derived by Wesnousky (1986) which relates maximum magnitude on a fault segment to its length is:

$$1.5M_w = 7.16 + 1.75 \log(1), \qquad 1$$

where l is the fault length in kilometers and M_w is the magnitude. For a given fault and given observation point, the Joyner-Boore equation [Joyner and Fumal, 1985] is used to compute peak horizontal ground acceleration from the maximum magnitude (M_w) and distance (d) to the fault:

$$\log(a_{max}) = 0.49 + 0.23*(M_w - 6.0)$$
$$-\log\sqrt{(d^2 + h^2)} - 0.0027*\sqrt{(d^2 + h\ 2)}, \qquad 2$$

where h is the average hypocentral depth (8.0 km for this study) and a_{max} is in fractions of the earth's gravity (g). Computation of spatial quantities such as fault segment length and minimum distance between a point and a fault is automatic in a GIS, and makes the calculations in (1) and (2) simple.

Slip rates

The fault locations and lengths yield maximum magnitudes and estimates of ground accelerations. However, a key ingredient in a probabilistic hazard map is the probability that a particular fault will rupture. The slip rate provides an estimate of a fault's activity, which can then be converted to a probability. Wesnousky (1986) showed that the slip rate can be converted to a recurrence interval by first estimating the seismic moment released and then calculating the slip in the maximum magnitude earthquake. The slip per maximum magnitude earthquake divided by the slip rate yields a recurrence rate for a particular fault. The probability that the maximum magnitude earthquake will occur during a time interval of Y years is given by:

$$P = 1 - e^{(-Y/RI)}, \qquad 3$$

where RI = recurrence interval.

As in the case of the fault segmentation, multiple tables of slip rates are available. We chose to use the compilation by Wesnousky (1986) because it was consistent with our fault segmentation and because it had been peer-reviewed. As will be discussed below, the fault segmentation and slip rates can be updated from other sources easily.

Computation of probabilities

The cumulative probability that a given point will experience a certain level of ground acceleration is the sum of the probabilities from all faults capable of affecting that point (Wesnousky, 1986). A GIS facilitates this summation because each fault can be buffered with a region within which the specified level of acceleration is exceeded (Figure 3). This region is assigned a probability from (3), and cumulative probabilities are computed by simply intersecting multiple regions (Figure 3). In this manner, a probabilistic hazard map is generated (Figure 4).

ADVANTAGES AND DISADVANTAGES OF A GIS

A GIS offers several advantages for maintenance of data bases as well as for computation of maps. Data bases can be readily updated, and multiple copies can be generated easily. The sensitivity of the final result to the component data can be tested. A GIS provides a standard format for exchange of data bases. Finally, comparison of the product (the probabilistic hazard map) to other data sets such as roads or utilities is easy.

Many of these advantages are also possible disadvantages. Easy duplication and modification can lead to multiple different copies of a data base, making preservation of a master copy a critical step. Overlays of product maps on other digital data bases can lead to misuse, especially if the question of appropriate scale is not considered carefully. Other disadvantages include user ignorance in GIS, lack of an established review process for data bases, lack of appropriate forum for "publication" of data bases, overload of libraries and other archival institutions with data bases, and lack of protection of intellectual property.

Updates and modifications

Publication and updates of analog geologic maps have been arduous and expensive processes in the past, resulting in long gaps in time between editions. A digital data base, on the other hand, can be updated readily and made available as soon as it has been reviewed for accuracy. For example, the fault data base described above was created two years ago and is currently being updated with new results for all of San Bernardino County. Additionally, slip rates and fault segmentation are being modified with new estimates from the SCEC Phase II report on the effects of Landers on southern California earthquake probabilities. This report has not yet been released, but the revised data base will be available almost as soon as is the report.

The ease with which modifications can be made may also lead to a proliferation of altered copies of the data base, however. Different schemes for fault segmentation or alternate slip rates can be substituted, leading to multiple versions of the data base. Ideally, users would document clearly the changes made but this rarely happens. A master copy of the unaltered data base will need to be stored in a universally accessible location to ensure that users can access the original data.

GIS literacy in users

The question of what constitutes the data base also arises. In the example presented here, is it just the fault data base with its locations, segmentation, and slip rates? Does the probabilistic hazard map in Figure 4 constitute another data base or is it part of the fault data base? The probabilistic hazard map in Figure 4 was generated for the Chancellor at UC Riverside to assist in the siting of an experimental reactor; hence the choice of 0.2g for acceleration. Another user may not be interested in that level of acceleration but may want a map for 0.4g. Do both of these maps then become archived? We propose that, rather than archiving either map in a data base, the algorithm developed from Wesnouksy (1986) be provided with the fault data base and the user be given instructions on how to generate the map needed. This minimizes the storage required for the data base, but requires that the user have some familiarity with GIS. Most users are currently ignorant of GIS however, and rely on GIS experts. This situation is similar to the state of scientific programmming in the 1960's and early 1970's. Most geoscientists today are capable of programming in a language such as FORTRAN, and the need for programmers to write programs for us is past. In the future, we expect that GIS will be simply another skill that geoscientists acquire in their undergraduate education.

Data base publication

There are two main obstacles to the publication of data bases: lack of a review procedure such as the peer-review process; and lack of a forum for publication. At SCEC, we have proposed a procedure wherein two independent scientists with GIS skills will be asked to review a particular data base. The review will include spatial accuracy, accuracy of tabulated data (attributes of geographic features), and completeness of documentation. We intend to submit the revised fault data base to this review. The lack of a forum for publication of data bases presents a more serious problem. At present, data bases such as the earthquake catalogs maintained by the U.S. Geological Survey or the California Institute of Technology are simply made available via Internet. Some organizations such as the U.S. Geological Survey or the California Division of Mines and Geology also distribute data bases compiled by their own scientists on CDROM. However, there appears to be no independent forum for data base publication.

This lack of a forum presents problems regarding intellectual property. Compilation of a data base is a time-consuming task for which the creators would desire proper credit. Upon distribution, data bases are labeled with the authors' names as well as the limitations of the data. While data bases can be copyrighted, the copyright information can be easily stripped off the electronic data. The data base can then be redistributed without proper credit, constituting a form of plagiarism. This lack of forum for publication also plagues honest scientists because properly citing a data base is then difficult. Our hope is that, as the publishing world moves increasingly to electronic publications, these difficulties will be eliminated.

We are using this proceedings volume as a forum to publish the digital fault data base. The Appendix contains instructions for how to obtain a copy of the data. In this manner, we are establishing a formal reference for the fault data base. This ad hoc approach is not the optimal solution, however.

Sensitivity studies

The probabilistic hazard map in Figure 4 is the result of combining several different types of data through an algorithm. Each of these data and each step of the algorithm has associated errors. Generation of one map in the analog world was so difficult that quantitatively analyzing the effect of each error on the final result was rarely done. Computation of the hazard within a GIS makes this step relatively easy. Automatic functions for computing spatial statistics such as percentage areas allows the user to determine the uncertainties in the final map and what effects various errors have on the map. A thorough analysis of Figure 4 is beyond the scope of this manuscript, but we provide one such example. Equation (2) has an error of ± 0.28 in the logarithm of the acceleration (Joyner and Fumal, 1985) which translates to a factor of 1.9 uncertainty in

TABLE 2. Effect of acceleration uncertainty on percentage areas.

ACCELERATION	1-20%	21-40%	41-60%	61-80%	81-100%
EQ (2) / 1.9	21.94%	22.28%	15.76%	17.27%	2.67%
EQ (2)	31.83%	16.36%	5.10%	0.76%	0.07%
EQ (2) * 1.9	25.50%	5.80%	0.44%	0.38%	0.00%

the acceleration. Table 2 summarizes the effect this uncertainty has on the percentages of southern California at each probability level for an acceleration of 0.2g. This uncertainty in the acceleration equation significantly affects the hazard estimate in southern California.

Comparison of data sets

The most powerful advantage of a GIS is in its ability to compare different data sets. As an example, Figure 5 shows a portion of the hazard map for San Bernardino superimposed on the network of roads. Such a comparison could allow a city or an emergency response team to plan for disaster response, as well as guiding use of resources in seismic upgrades.

Misuse of data bases

The author of a digital data base is responsible for accurately and thoroughly documenting how the data have been entered, the sources of the data, the scale used, and the limitations of the data. As mentioned above however, all of this information can be removed or ignored by the user. The biggest hazard for misuse is in expansion of the map to a scale that is invalid for the data. For example, the fault data base has been compiled at a scale of 1:24,000 with points every 100 m along the rupture zones. While an appropriate use would be to determine if a neighborhood lies in a rupture zone, attempts to determine if a rupture crossed an individual plot would be inappropriate. The author of an analog map now has no control over the use and misuse of the data by users, and it is unrealistic to assume that the author of a digital data base will be able to control its use either.

CONCLUSIONS

The proliferation of digital data bases in the geosciences is accompanied by both advantages and disadvantages. Problems include modified copies of data bases, protection of intellectual property, user illiteracy in GIS, lack of a publication forum for data bases, electronic storage requirements, and misuse by others. Significant advantages result from the use of digital data bases which override the possible problems, however. Digital data are easy to update, resulting in more current information. Creation of digital data bases within a GIS permits easy exchange of data and use of automatic functions for spatial analysis. For the researcher, the ability to search for patterns between disparate data sets represents the most significant advantage. For both the user and the research scientist, the ability to compare the researcher's results to cultural data such as roads and utilities is a crucial advance.

APPENDIX

The digital fault data base is available from SCEC via anonymous ftp. The data base is in ARC/INFO export format and can be copied with the following steps:

```
ftp vortex.ucr.edu
login: anonymous
password: {enter your e-mail address here}
cd scec
binary
get alqfltall.e00
bye
```

The file must then be imported into ARC/INFO or another GIS.

REFERENCES

Aki, K., and Richards, P.G., 1980, Quantitative Seismology, volume I: San Francisco, California, Freeman and Company, 559 p.

Clarke, M.M., and others, 1984, Preliminary slip-rate table and map of late Quaternary faults of California, U.S. Geological Survey Open-file Report 84-106, 13 p.

Hart, E.W., 1992, Fault-rupture hazards in California, California Division of Mines and Geology Special Publication 42, 32 p.

Jennings, C.W., 1975, Fault map of California with locations of volcanoes, thermal springs, and thermal wells: Scale 1:750,000 Sacramento, California, California Division of Mines and Geology Geologic Data Map Series no. 1.

Joyner, W.B., and Fumal, T.E., 1985, Predictive mapping of earthquake ground motion, in Ziony,

J.I., ed., Evaluating Earthquake Hazards in the Los Angeles Region — An Earth-Science Perspective, U.S. Geological Survey Professional Paper 1360, p. 203-220.

Rogers, T.H., 1965, Geologic map of California, Santa Ana 1:250,000 sheet: Sacramento, California, California Division of Mines and Geology.

Ward, S.N., 1994, A multidisciplinary approach to seismic hazard in southern California, Seismological Society of America Bulletin v. 84, p. 1293-1309.

Wesnousky, S.G., 1986, Earthquakes, Quaternary faults, and seismic hazard in California: Journal of Geophysical Research, v. 91, p. 12,587-12,631.

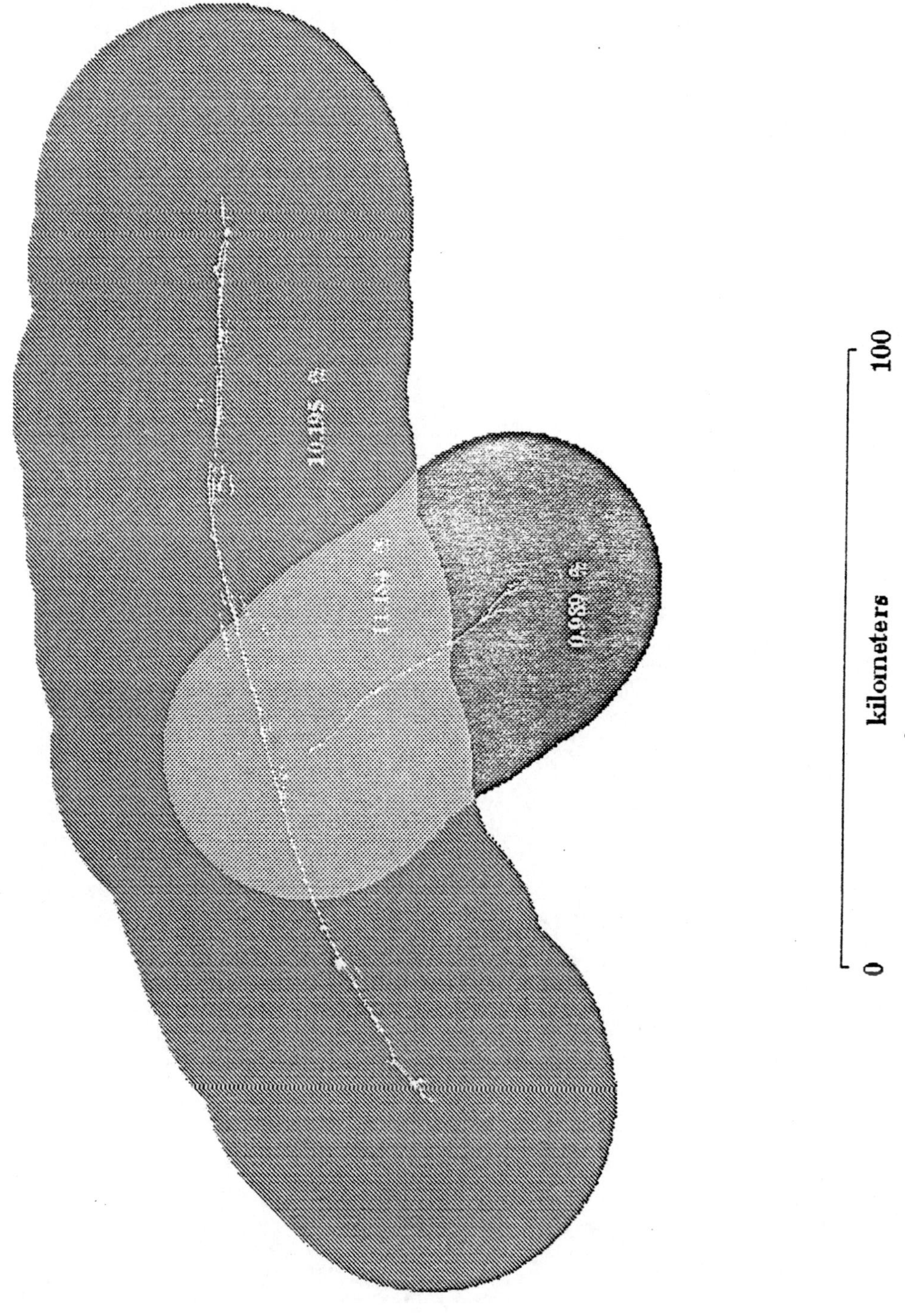

Figure 3. Faults surrounded by regions which would experience acceleration greater than 0.2g. Each fault has a probability of that acceleration occurring in 50 years, and the intersection of the two regions has a probability equal to the sum of the individual contributions.

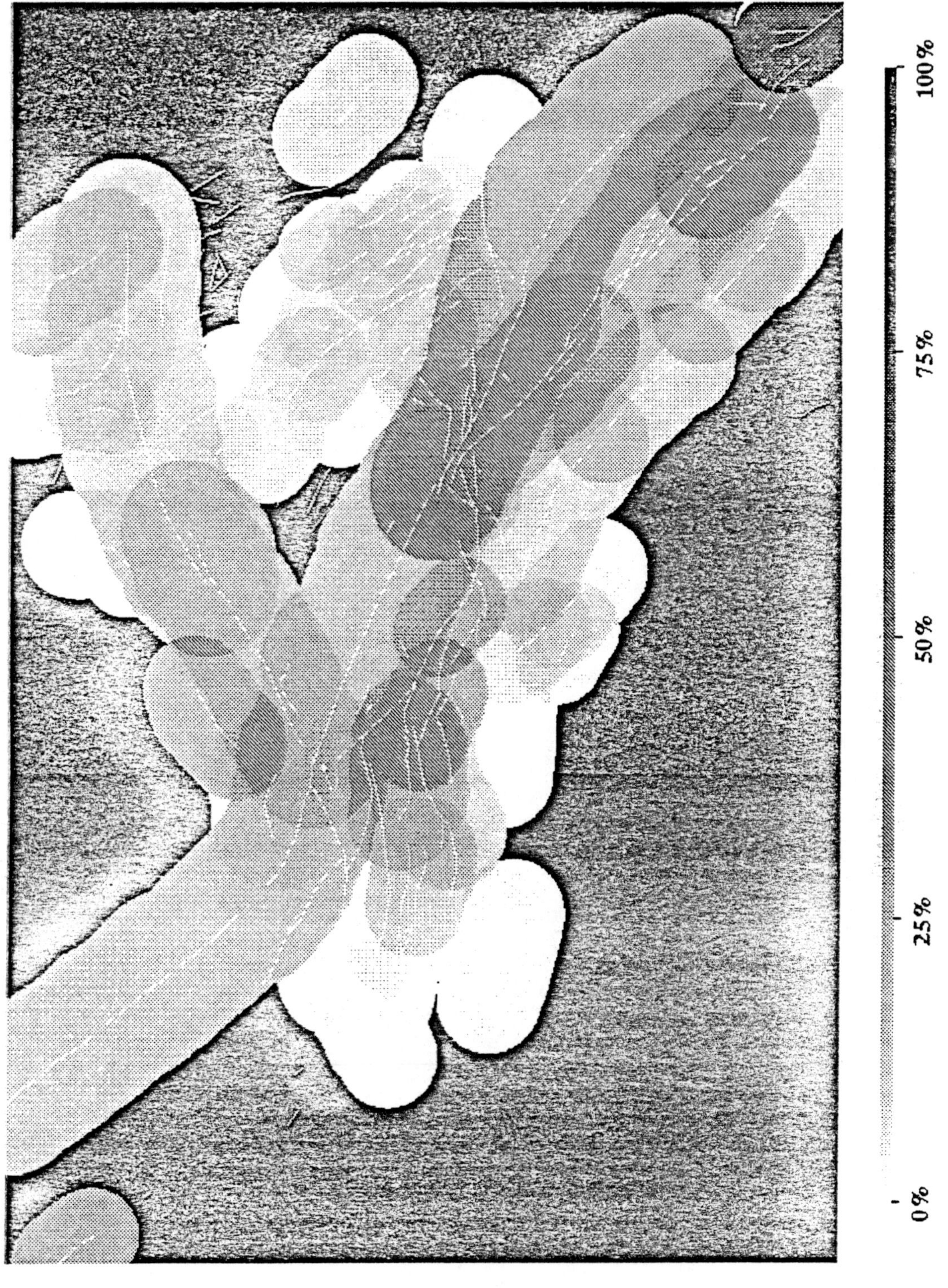

Figure 4. Probabilistic hazard map for acceleration of 0.2g in the next 50 years.

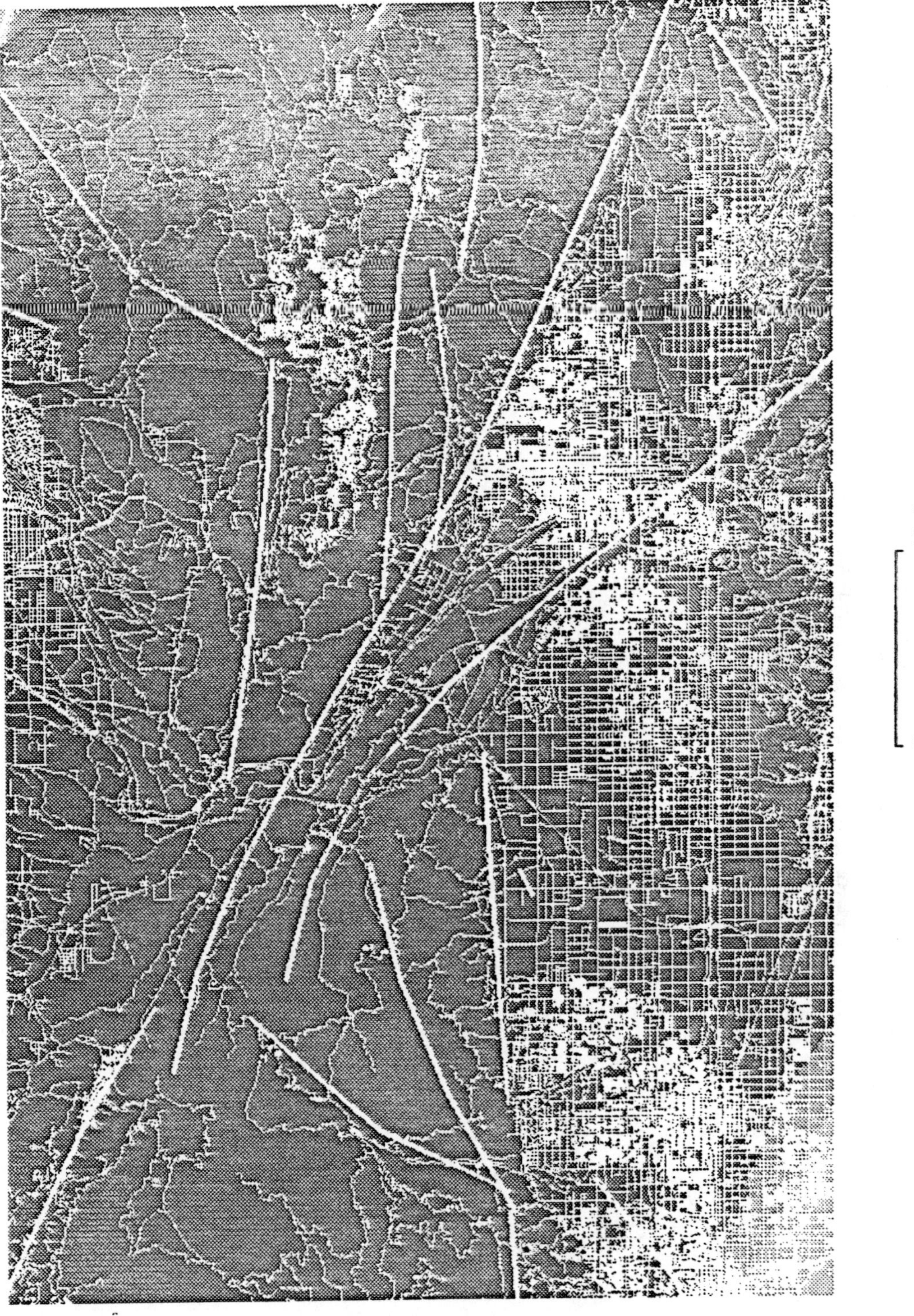

Figure 5. Comparison of probabilistic ground shaking from Figure 4 with road network for the San Bernardino Basin.

THE NATIONAL PERFORMANCE REVIEW:
A REINVENTION LABORATORY AT THE U.S. GEOLOGICAL SURVEY –
THE FUTURE OF INFORMATION AND PRODUCT DISTRIBUTION

Wendy R. Hassibe

U.S. Geological Survey, National Mapping Division
RMMC Bldg. 25, MS 508 Federal Center
Denver, Colorado 80225

Abstract – For many years, the distribution activity of the U.S. Geological Survey has been encumbered by an inadequate computer system, paper-based, outmoded processes, cumbersome financial regulations, an inflexible work force restricted by classifications, inadequate product information and the inability to deliver products to customers in a timely manner.

The National Performance Review is an effort sponsored by the Vice President of the United States to provide for a more responsive, customer-focused federal government. A series of Reinvention Labs were established in July 1993 to begin specific projects identified by individual agencies. The Reinvention Lab team at the U.S. Geological Survey was comprised of representatives from all parts and levels of the organization from warehouse worker to program managers. The report prepared by this Team was accepted, and implementation is now underway.

The new Information Dissemination System focuses on todays business practices, incorporates technologies including product identification, electronic ordering and a new approach to providing for greatly improved customer access and response to their needs.

EXECUTIVE SUMMARY

Vice President Gore's National Performance Review (NPR) of the Federal Government was established to "reinvent systems of government, redesign agencies and programs to make them more responsive to their customers, and streamline the government" (National Performance Review (U.S.), 1993.)

Seven Reinvention Laboratory Teams at the Department of the Interior were approved, each given the task of reinventing a bureau unit or process into a streamlined customer-oriented organization. The teams met together in Columbus, Ohio for two weeks of training on organizational reinvention and reengineering concepts. The teams were instructed to use these and other concepts to create a "clean plate" design of their process. Designs were to be data-driven and reflect extensive research by the Teams.

The U.S. Geological Survey (USGS) Team focused on information and product distribution within the agency. The seven-member Team is made up of six USGS employees and one member from the National Oceanic and Atmospheric Administration. Based on the training, one of the first Team actions was to develop a shared mission statement:

To develop an earth science information

dissemination approach which is Customer-driven and continuously provides superior service.

Our research included interviews with ten private sector organizations known for their distribution capabilities; interviews with stakeholders from USGS; meetings and open discussions with employees of the distribution and information activities and extensive reviews of numerous documents ranging from previous studies to customer surveys, to guidelines published by the American Warehousing Association.

Research revealed an existing system that was rife with problems ranging from the inability to obtain fundamental management information, to long turnaround times for filling orders. This research also corroborated one of the fundamental premises of the National Performance Review: it's the system that is a problem, not the people.   Consistently, we found dedicated employees working beyond expectations to serve their customers, usually in spite of the system.

To call the distribution system "a system" is really a misnomer. One of the basic problems identified was no current systematic design for either the information process as a whole or the distribution function within it. Interviews clearly indicated a strong organizational bias towards science which partially explained why

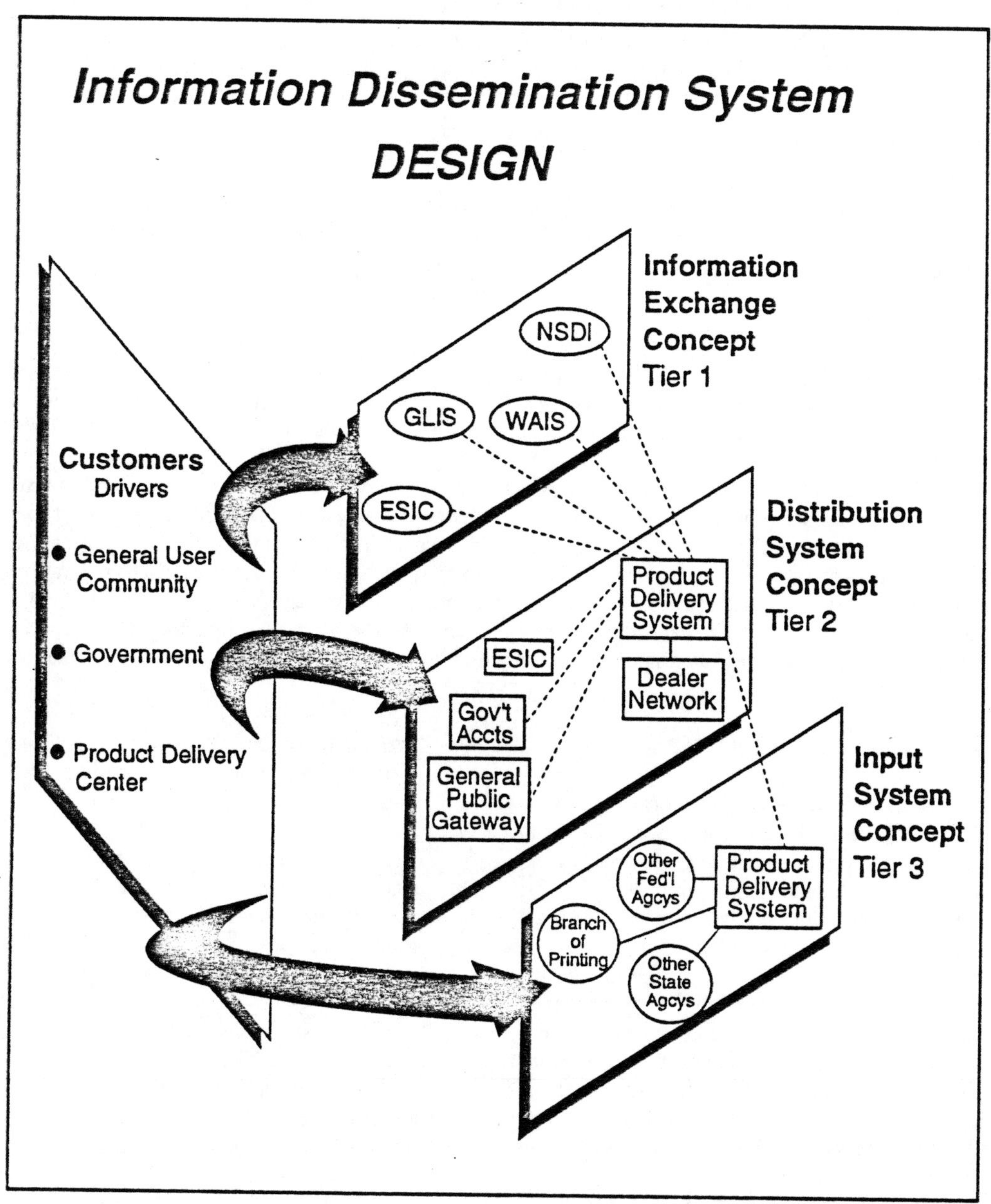

Figure 1. The Information Dissemination System Design of three component tiers addresses how all customers receive information about products (Tier 1), how those products are distributed to the customer (Tier 2), and how those products enter the distribution mechanism (Tier 3).

distribution and information issues historically have taken a back seat to "core" scientific programs, especially with regard to budget and technology.

Critical issues regarding the current distribution system include:

- Computer system does not support effective management of the activity.
- Fiscal, personnel and management processes are encumbered with real and perceived rules and processes.
- Quality customer service is not sustainable.
- Culture of the organization places little or no value on the operation.
- Lack of substantive communication at almost all levels.
- Employees, at all levels, cannot "do their jobs" because they are encumbered by the processes, rules and procedures
- Existing industry standard practices are not used.

In response to these issues, and following extensive data collection, the Team developed a three-tiered concept for the Information Dissemination System. This approach recognizes that distribution does not occur in a vacuum but is an integral part of the larger information macrocosm.

The tiered design of the system, depicted in Figure 1, portrays the interrelated nature of the three components:

- Tier 1 Information Exchange
- Tier 2 Distribution System
- Tier 3 Input System

Both Tier 1 and Tier 3 are critical to the overall design of the Information Dissemination System. One provides the information about the products; the other supplies the products. Although our Team focused on product distribution and did not analyze Tiers 1 and 3 in depth, this report does contain recommendations for reengineering these two components. This analysis will be critical to full implementation of the concept.

DESCRIPTION OF TIERS

Tier 1: Information Exchange

This includes utilization of wide-reaching electronic data bases and networks such as the Wide Area Information System (WAIS), Global Land Information System (GLIS) and the Network of State Earth Science Information Centers (ESIC), via INTERNET. Scientific information is exchanged, customer inquiries are responded to and data is transferred from site to site. Online distribution of data can occur within this framework when that approach is fully developed.

Tier 3: Input System

This includes the sources of products such as the Branch of Printing and other federal agencies.

Tier 2: Distribution System

The primary focus of this report. Here, the physical realities of getting a product into the hands of a customer becomes the driving force. The heart of this design is the Product Delivery System, which is made up of three Teams:

- Customer Fulfillment Team
- Systems Management Team
- Management Team

Customer Fulfillment Teams are responsible for their customer from start to finish. They are empowered, cross-functional and can handle processes including customer account services and materials handling. These teams are assigned to blocks of customers, such as: federal and state, ESIC's, Map Dealers and the contractor operation (General Public Gateway).

The Systems Management Team is also empowered, cross-functional and responsible for maintaining and improving the technological systems that are the backbone of the concept. This Team would also be responsible for information data bases, interactive catalogs and communication linkages within all three Tiers.

Management Team is empowered and serves in a support function for policy guidance, planning, budgeting and human resource management.

All three Teams have performance measures based on outcomes and are pilot applications of pay banding.

The Distribution System is linked via Electronic Data Interface (EDI) to:

- Federal and state customers
- Contractor (General Public Gateway)
- USGS Earth Science Information Centers
- Map Dealer Network

EDI is the technology that provides access to the products. Anyone can gain direct access to the distribution system via EDI if they so choose. This option will be offered on a subscription-type basis, and customer groups using this access will be provided a discount. This approach virtually eliminates the order entry function in the existing system. Customers without EDI access, chiefly the general public purchasing 1 or 2 products at a time, will be served primarily by a contractor. This contract, referred to as the General Public Gateway, will process customer requests that come in on a 1-800 line, by mail or FAX. The contractor will provide for credit card purchases, purchase orders and checks, plus options to overnight delivery at appropriate charges. The USGS ESIC's and the Map Dealer Network will also be responding to customers without EDI access and these three components (contractor + ESICs + Dealers) will be linked electronically in order to refer customers to the appropriate site. In addition, the newly-defined Map Dealer Network (using EDI) will be encouraged to establish their own customer-supplier network to provide products to smaller, low-volume dealers who do not have EDI access.

Along with EDI, the other critical technology required for this concept is bar coding. While the Team realizes that it will take time to bar code all USGS products, the sooner a full scale conversion to bar coding occurs, the easier it will be to implement this design.

The key measurement of success for the distribution of products is "Right Product Right Time," or RPRT. Quantifications of this measurement focus on outcomes, rather than outputs -- did the customers receive what they wanted, when they wanted it, rather than how many maps are shipped in a day. Another measurement will be a Board of Customers to be supported by the Management Team. The Board will represent customer groups and include representatives from the private sector.

As with anything new, change will need to occur at several levels. Information and product distribution must be recognized as a critical component of our organization. Traditional pay and classification structures must evolve into more flexible paybands and elimination or easing of regulations restricting customer surveys must take place. The Vice Presidential report includes a recommended action (that) "the Office of Management and Budget will delegate its survey approval authority under the Paperwork Reduction Act to departments that are able to comply with the act."

The impact of full implementation will result in possible cost savings through the elimination of order-entry, virtual elimination of mail and funds handling and a reduction in turnaround time from several weeks to a matter of days.

Actions needed to prepare personnel for these changes include an implementation plan, plus a series of workshops involving senior management and NPR Team members. Plans for implementation are based on creation of a full-time team that represents all levels of the activities and will focus 100% of their time on development of the plan and requirements. The workshops will explain the concepts and application of reengineering and the specific System concept. Reengineering of components in Tiers 1 and 3 needs to begin at the same time. Teams should be identified and trained immediately. Full implementation of the System concept should occur within 2-3 years.

When fully implemented, the proposed concept will be capable of handling expanded product lines from throughout the Department of the Interior and other departments, which could result in additional revenues for re-investing in capital expenditures and expanded information capabilities. The key to the concept's success is the commitment to the customer.

SCOPE OF REVIEW -- BACKGROUND

Authorization for the dissemination of information and products by the USGS is based on Title 23, U.S. Code, Section 42:

> The Director of the Geological Survey is authorized and directed, on the approval of the Secretary of the Interior, to dispose of the topographic and geologic maps and atlases of the United States, made and published by the Geological Survey, at such prices and under such regulations as may from time to time be fixed by him and approved by the Secretary of the Interior.

Sub-sections of the same Title provide legal authority to distribute and sell earth science information, including published maps, aerial photographs and other remotely-sensed products, and digital data. In addition to the legal authority, information dissemination is recognized as part of the USGS mission.

The mission statement of the U.S. Geological Survey is:

> to provide geologic, topographic, and hydrologic information that contributes to the wise management of the Nation's natural resources and that promotes the health, safety, and well-being of the people. This information consists of the water, energy, and mineral resources, land surfaces underlying geologic structures, and the dynamic processes of the earth" (U.S. Geological Survey, 1985).

Based on these authorities and directives from the Department of the Interior National Performance Review training program, our team established this mission statement, "To develop an earth-science information dissemination approach which is customer-driven and continuously provides superior service."

CONCLUSIONS

This Review was proposed in response to problems and issues raised over several years by extensive management studies, by employees and by regulatory office such as the Department of the Interior (DOI), Office of Inspector General; DOI Office of Financial Management (Management Control Review), and the USGS Office of Financial Management. Critical concerns included:

The computer system(s) does not support effective management of the inventory, ordering and sales of products, customer needs, budget requirements, costs of operation and staffing requirements.

Fiscal, personnel and management processes prevent, or are perceived to prevent, accomplishing the mission. For example, personnel regulations or processes prevent responding to seasonal workload demands; budget regulations or processes prevent the use of revenues for infrastructure, long-term systems planning, facilities improvements and performance incentives.

Inability to sustain even good customer service, let alone excellent customer service.

The organizational culture does not see significant value in the information and distribution functions. At least 18 studies have been performed and recommendations made to management. While some recommendations have been implemented, the ability to distribute earth science information and products effectively continues to decline.

Changes since August 1994

Since mid-August 1994 we have implemented:

- Reduced internal turn-around time to 7 days. It was at 10 weeks in August.
- Established ½ of our workforce in cross-functional, flexible and generic position descriptions.
- Translation: everyone can work everywhere to deliver products to customers.
- We are working on implementation of credit cards for all customers, in addition to electronic ordering of products for dealers.
- We have a "horizontal" organization — one supervisor for each 50 employees.
- We have entirely changed our correspondence — customers are now being called directly (instead of writing letters), more flexible billing, more pro-active customer service.

We expect full implementation within a year, but our progress will be ongoing. This is not a start-and-stop process, but really more about how we should be doing business.

REFERENCES

National Performance Review (U.S.), 1993, Creating a government that works better & costs less: Report of the National Performance Review: Washington, D.C., Government Printing Office, 168 p.

U.S. Geological Survey, 1985, The goals of the U.S. Geology Survey: U.S. Geological Survey Circular 1010, 17 p.

PUBLISHING IN A DIGITAL WORLD:
THE PAST IS THE KEY TO THE FUTURE

John M. Aaron

U.S. Geological Survey, 904 National Center
Reston, Virginia 22092

Abstract — Publishing today stands poised between two revolutions: the first, printing, is over 500 years old and nearly spent; the second, digital electronic, is little more than a decade old and barely started. Although the technologic drivers are radically different, the two revolutions share many characteristics that provide some important clues about what to expect from publishing in a digital world. And, although it is much too early to predict with any confidence exactly where electronic publishing will take us, it is certain that the nature of publishing, communication, and information dissemination will change fundamentally. The change will have profound impacts on society that will equal or exceed those that followed the shift from script to print after the invention of the printing press and movable type.

Major changes to expect (many are already underway) in publishing ventures include a shift from paper products to digital products; increased emphasis on the nature of digital information as a valuable, strategic asset; greatly increased emphasis on content and the needs of end users; a large-scale redefinition of publishing processes and relationships; and a merging of the enabling technologies that guide publishing, communications, and entertainment and, hence, a blurring of the distinctions between these enterprises. These and many other changes will be linked to the evolution of computer and related technologies, but technology itself only provides the foundation. The greatest challenge will lie in knowing what to do rather than in knowing how to do it.

PUBLISHING IN A DIGITAL WORLD

The theme of this symposium is "Changing Gateways: The Impact of Technology on Geoscience Information Exchange." Surprisingly, the change in the geoscience publishing environment is slow and limited; geoscience publishing today looks pretty much as it did a century ago. Can we predict what it might look like a century from now? Not precisely, of course, but we do know that fundamental change is always much slower and, in the long term, much more profound than we imagine.

To better understand this assertion, let us look at ourselves as a professional community of earth scientists and earth science information experts. Despite a mountain of evidence that there is a technological revolution going on around us, the premise of this symposium, the publishing culture of the geological sciences and most other scholarly disciplines is still very traditional. Maybe it is because we are so accustomed to thinking of change on the scale of geologic time that we are also comfortable with the glacial pace of change in our traditions, institutions, and processes. Or maybe we just like things the way they are now, pretty much the way they have been for the past couple of hundred years. Nevertheless, we are about to witness a paradigm shift as profound as anything that plate tectonics did to

geology, Darwin did to biology, and Gutenberg did to printing—the world of electronic publishing.

Electronic publishing is defined not only by the computer hardware and software tools used to create, format, and illustrate scientific reports and generate camera copy for printing, but also by the electronic communication of those reports to intended readers worldwide. The link between author and audience (with or without organizational sanction; perhaps untouched by peer reviewers or an editor) is the world's largest publishing and communications machine—the Internet.

What is the allure of electronic publishing? The advantages over paper are, and have been, obvious for well over a decade. They include:

- Faster and wider access to information

- Information that is more easily kept up-to-the-minute

- Much more flexibility in the way consumers can use information

- Ready access to powerful information processing tools

- Much easier integration of electronic data with myriad applications

- Reduced cost of wide dissemination of information.

The implications of these advantages are now starting to take hold. After a long period of gestation in which people have struggled to understand the opportunities, the promise, and the problems offered by electronic publishing and telecommunications technology, conventional wisdom has finally started to shift. The pressure to change is building, and the change in outlook is palpable. Publishing enterprises are now on the verge of a major transformation, and the reasons seem pretty clear:

- Researchers and academicians want to take advantage of the speed and communication advantages offered by electronic publishing technology. Why target your work to 2000, 5000, or even 10,000 people in North America when you could potentially reach many times that number around the world? And why use conventional postal services when electronic networks could reach your audience in a few seconds or a few minutes?

- Commercial publishers increasingly believe that higher profits will flow from new electronic products or electronic products that supplement their present line of paper publications. (The commercial sector is now the fastest growing part of the Internet.)

- Many professional societies are looking for new ways to serve their members or, in some cases, ways to try retain their current members.

- Libraries surely are aggressively looking to improve services and save resources at the same time. These are institutions under siege. For many of them, the switch to electronic systems and saving money is a matter of survival.

- And finally, the Federal Government is making electronic publishing easier by subsidizing electronic networks and funding demonstration projects. The Internet itself grew from a Federal Government (Defense Department) enterprise, and to most researchers it still appears to be free.

If the advantages of electronic publishing are so patently obvious, where is the resistance and why is it taking so long to change how we publish and distribute data and information? For the answers, we can start by looking at the professional culture of the geological community. Our publishing traditions are mostly derived from our journal culture, which remains traditional, and this certainly applies to the academic world as well as to the USGS and similar organizations. We all tend to operate a highly formalized publication process that rests on the concept of peer review.

Scholarly articles and reports are not accepted for publication outright but are circulated among the author's peers and many times reworked and revised to accommodate reviewers' changes, and perhaps revised again to accommodate editorial changes. In the electronic world, in some disciplines, challenges to this "business as usual" are already occurring in the form of direct, online communication from authors to peers, perhaps untouched by reviewers, editors, or society sanctions. The freedom, lightning-fast results, and apparent lack of cost lull all into believing that this is a good thing. However, such a system can also be a recipe for chaos and confusion if we are not careful. The checks and balances that characterize the winnowing process of the traditional publication system serve an essential role in selecting, authenticating, and disseminating sound data and information based on credible science. The many public benefits afforded by this accrediting process could be lost in an electronic free-for-all.

Another characteristic of our journal culture that results in resistance to change is the informal alliance that exists among the key players: authors, publishers, libraries, and government. Authors supply scholarly articles, and their professional reputations and career advancement depend upon getting the articles published. Professional societies, government agencies, and commercial enterprises publish the articles supplied by authors. Libraries are probably the largest market for what gets published, and government directly or indirectly subsidizes the entire process through research grants, payment of page charges, and other means. No single overseer is in charge of the entire publication process. The stakeholders all have different interests, and none is in a position to change the process or the relationships very quickly.

Besides the culture, especially the fact that no single entity is in charge of the scholarly publication process, several other points of resistance come to mind:

- The printed page has been paramount. It will be a very slow process, I think, to wean both authors and readers (above the age of about 16) away from the familiar page paradigm and the inherent control of the sequence and pace at which they present or access the contents of a report or article.

- Changeover costs from traditional to electronic publishing could be high. Who will pay them?

- Many people and organizations automatically adopt a wait-and-see approach to any new enterprise. It is safer than gambling on a course of action, the outcome of which is not entirely clear.

- A huge, huge impediment to change is the reward system that exists in both academia and the research environments of the USGS and similar organizations: make no mistake, it is still strongly geared to paper publications. People in virtually all scholarly disciplines, including computer science, perceive that little credit is attached to publishing electronically, even in formally established online journals in disciplines in which they exist. The medium has not yet gained the confidence of potential users.

There are also several unresolved rather practical technical and business questions and problems that come to mind. For example:

- How can unlimited copying be prevented in the electronic environment? Once a paper "goes digital," there is virtually no limit on where it can be sent or how many times it can be copied or altered, which can violate the copyright and threaten the integrity and income of authors and publishing organizations.

- Can the electronic journal generate enough production cost savings and additional income to compensate publishers for subscription revenues lost from their print products? Or will the revenues not really be lost but merely redirected to electronic media?

- How do we deal with fonts, special characters, and graphical elements? Is plain vanilla ascii sufficient? And how does the electronic environment deal with richly formatted documents? Right now not very well. Another significant consideration is that presently the majority of the world's computers are *not* on the Internet, and of those that are on the net more than half are not equipped to read files that are not in ascii or their native word processor formats.

- What is the "lowest common denominator" computer platform, video resolution, and software user interface? These change constantly, unlike the universality of ink-on-paper, a crash-proof, higher-resolution, more convenient user interface.

- Retrospective conversion is an important issue for libraries. A single year of a journal online or on CD is nice, but is that enough? Once users make the commitment to electronic products, they will want electronic access to back issues as well. How far back should publishers expect to go in converting print products to electronic media? At what cost? And who will pay for it?

- Bandwidth: how much traffic can the electronic superhighway bear? Will electronic transmission of large, complex documents be like trying to force watermelons through a firehose?

These cultural, technical, and business issues can be daunting and explain, in part, why the move from print to electronic publishing is so slow. Yet, you need not wonder if it will ever happen. The full shelves and floors piled high with books and magazines that you find in your newsstands and bookstores and your own bookshelves are the terminal moraine of a technological revolution that began over 500 years ago and now is nearly spent. Electronic publishing is the leading edge of a new revolution that is barely started, one that we do not yet fully understand in terms of where it will lead us, but one about which some things can already be said with certainty.

First: The future will be digital. In the developed countries, we are rapidly creating an infrastructure in which most information and entertainment will be prepared and delivered in digital form. With text and sound, we are already mostly there. Computer word processors and compact discs have replaced typewriters and vinyl phonograph records. In both cases, the tools of creation and the products are digital. The graphic arts are not far behind, moving in the same digital direction by developing powerful illustration and data visualization capabilities that far outstrip anything possible in the print world.

Second: In a digital world, everything changes (and, for that matter, everything is changeable). Everyone involved in publishing, telecommunications, entertainment, and computers will be affected. All of our jobs are going to change in some measure; some have already. What we do in one area of endeavor, such as publishing, will both affect and be affected by what happens in other areas because we are developing a common digital infrastructure upon which all our work is based and interconnected. And all of us are potential customers for the services and products that will emerge in this environment. This is why phone companies and cable companies are battling each other so fiercely for the right to deliver products (information, entertainment, and services) to television or computer screens where you live and work. The financial stakes are enormous, and so are the benefits.

Third: Anyone who has followed the fortunes of publishing technologies over the past decade is acutely aware of how much some things have already changed. Our technology base has evolved away from highly specialized proprietary publishing systems **to** mainstream desktop computer and telecommunications applications. In fact, much of the growth and pizazz of desktop computing is related to the publishing and communications applications that were developed for that environment: word processors, graphics tools, page description languages and page make-up tools, virtual document communicators, compact disc technology, the rapidly growing suite of online publishing tools, and, of course multimedia (which one industry observer has characterized as "multimediocrity" because, so far, most implementations of it have been primitive and uninteresting). In the span of a few years, publishing has changed from a craft to a computer application that is no longer bound by time, place, medium, view, or source. The coming generation of publishing and information technologies will irrevocably change the way that individuals and organizations acquire, manage, and communicate information. This means that we in publishing and information dissemination enterprises must now play in a much larger arena. In particular, we have to pay close attention to developments in computing and telecommunications and probably in other fields such as entertainment.

THE PAST IS THE KEY TO THE FUTURE

The kinds of changes that we are experiencing in publishing are not particularly new or uniquely characteristic of our time. In fact, many aspects of what is happening in publishing today broadly parallel the 70-odd year period following Johannes Gutenberg's synthesis, around the year 1440, of the printing press, movable metal type, and ink and paper. The spread and adoption of that new technology embody the beginnings of modern printing. Perhaps the most remarkable aspect of Gutenberg's achievement is that his printing system

was not significantly improved upon until the 19th century.

The following table summarizes the many enormous social, cultural, and economic changes that accompanied the shift from script to print in the several decades following Gutenberg's invention. As Table 1 cryptically shows, those changes were profound. The world was transformed. It was a revolution in every sense of the word. Obviously it did not happen overnight, and you can imagine that there were massive problems and disruptions along the way. The similarities between then and now are notable. Geologists are taught that the "present is the key to the past," but it is also true that the past is the key to the future. Some of the parallels between the printing revolution and the electronic revolution are truly astonishing, and they can help us to better understand and put into clearer perspective some of the changes occurring today in publishing enterprises. Many of the fears and concerns that troubled people trying to cope with changes in the late 15th century sound very familiar. Here are a few:

Information overload. Before the advent of the printing press, scholars had few books to master, usually only one — a religious tome. With the spread of the new printing technology, the task of keeping abreast of new printed works became oppressive. More was worse. It is not so different today, when we are literally drowning in data and information, much of it collected and processed electronically. The volume of data and information is growing exponentially. And now we have the Internet on which to move it all around the world in seconds. That is great, but keeping up with all of that data

TABLE 1. The First Printing Revolution.

A Shift from Script to Print	
From	*To*
Scribes	Print shop
Church control	Entrepreneur control
Slow and expensive	Fast and cheap
Corruption by manual copying	Preservation by printing
Unique copies	Identical copies
Oral transmission	Written transmission
Memory	Literacy
Narrowcast distribution	Broadcast distribution
Limited communication	Mass communication
Science: words	Science: images
Religion: images	Religion: words
Divine revelation	Logical reasoning
Private interest	Public good

and separating the useful from the irrelevant have become an enormous challenge, even for workers in small, specialized disciplines. It would appear that the "electronic superhighway" desperately needs some strategically sited information landfills where the electronic litter can be collected and buried.

Disenfranchised professionals. In the late 1400's the spread of printing technology made the role of scribes obsolete. They gradually disappeared from the workplace; they either learned new skills (becoming editors, typesetters, and librarians) or retired. We see the same thing happening today. For example, typesetting, an honorable craft, has been largely replaced by inexpensive software running on desktop computers. Although many other occupations usually associated with publishing enterprises will be rendered obsolete by advances in computer and telecommunications technology, old work will be replaced by new work, rather than no work. Continuous retraining in needed new skills will become routine in the workplace.

Intellectual property and copy rights. Who owns what rights to what works? Copyright law is as old as the Gutenberg press. It was contentious then, and it is no different today, though the protections afforded and the issues have changed over the years. Today, use rights and copyrights rank among the most confusing, quarrelsome, even threatening issues facing the originators, publishers, and users of creative intellectual works. As the debate moves from a focus on print products to electronic products, a huge new factor is the culture of the Internet, which aggressively promotes the *free* flow of information regardless of the source.

Aesthetic pollution. The advent of the printing press gradually transformed the workforce, bringing to it many new people who possessed no grounding in or notions of the accepted practices of the craft, design, and aesthetics of scribing. In many respects, it is no different today. Traditional typesetters, other artisans, and editorial and design professionals are being replaced by all manner of people operating desktop computers in all segments of the work environment. Unfortunately, computers possess no native design intelligence or good taste; that must come from the operator, but frequently it does not. The result is a growing collection of documents that traditional publishing professionals consider aesthetically unappealing and ineffective. I regularly see newsletters, even from within in my own organization, that are so typographically complex, unattractive, and uncommunicative that they resemble ransom notes.

But there is another subtle but important message to be gleaned from the history of the printing revolution: Gutenberg gave the world the technology to start the revolution, but he was not around for the fundamental, albeit slow, transformation.

The person who can be credited with fueling the revolution was a Venetian named Teobaldo Mannucci, who was known by the Latin name of Aldus Manutius. Today we associate his name with a software merchant (Aldus Pagemaker and other publishing applications), but, in his time, Aldus was a world-class entrepreneur. He founded the Aldine Press in Venice in 1494, 54 years after Gutenberg's inventions and 28 years after Gutenberg himself died a pauper. It was Aldus, his publishing enterprise, and the venture capitalists he attracted that really transformed and commercialized the printed word by producing Greek classics for the masses. Italic type was first designed and cut by an artisan working for Aldus, and Aldus is also responsible for many other typographic and printing innovations. By far his greatest contribution, however, was in making printed books portable and inexpensive.

Most of the seminal technologies that define our civilization and culture traveled a similar long path from invention to widespread adoption. In most cases, the inventors did not even dimly perceive the true importance of their creations; most, including Gutenberg, had something else in mind altogether. And in most cases it took 25 to 50 years, sometimes longer, for society to accept the new idea, understand and act on the opportunities, and turn the inventions into cultural icons. This was true of the Gutenberg printing press, the telephone, cinema, radio, the automobile, the airplane, and television. The history of the microcomputer will be no different but possibly more compressed. This revolution has barely begun; we are still in the incunabula.

It is certainly true that the evolution of technology-related change in our organizations, institutions, and work processes seems chaotic and unpredictable. It is also true that the new territory is mostly uncharted. The conventional wisdom blames technology for all of this, but technology itself is not really the culprit. It is not a magic wand that we can wave at our problems and expect them to be solved, nor does it constitute a toll-free superhighway between information creators and information consumers. Technology does not itself create or drive change. It merely serves as a catalyst for change by establishing a framework that opens up realms of new possibilities and provides new choices. Change comes as a result of how we choose to adopt and use new technology. There are many roads to the future, but not necessarily to the same future. Deciding *what* to do (our strategy for dealing with the available choices) will prove a lot more challenging than knowing *how* to do it (capabilities offered by technology).

I reiterate two important points about change noted above: it is painfully slow and mostly unpredictable. Why? Because it is absolutely controlled by human factors. Real, gut-wrenching fundamental change of the sort that we are wrestling with today in publishing and information dissemination enterprises requires people to break the intellectual and cultural bonds that tie them to the past and to things and ways that are familiar and comfortable. A huge amount of inertia must be overcome. Thus, change occurs not at the speed of thought, invention, or fiat but rather at the speed at which new ideas are disseminated, absorbed, and accepted, which is limited by the weight of habit and the willingness to accept and adapt to change. Abundant historical evidence indicates that this process requires at least a generation. The course of change is unpredictable because it is driven by choices and strategy, which are influenced by different stimuli in different people: logical, practical, emotional, financial, political, social, and cultural. In the long term, this game of chance is a random walk in which you cannot predict the final destination.

One last thought is a word of caution. New information and communications tools are seductive and, properly managed with sound strategies and insightful direction, they offer us wonderful opportunities to move aggressively toward fundamental changes in the systems and processes by which we organize, analyze, archive, and communicate geoscience information. However, as important and valuable as these tools are as shapers and communicators of information, they are utterly useless without the raw material of relevant content. *Content* takes primacy over the tools and, as the volume of information grows and clogs the lanes of the "information superhighway," *context* (content that is searched, filtered, linked to specific user needs, and presented in a form easily accessed and digested by the end user) will become a more valuable commodity than pure content.

We also need to bear in mind that although there are more than 25 million people using the Internet, and that number grows by about a million per month, the reality is that only a tiny fraction of them really care about the information produced by earth scientists. Not many people are going to read or download our data and reports just because they are there; in fact, the growing volume of data from myriad sources will discourage information "surfing" and put a premium on context.

In addition, many of us work in "high-tech" institutions where we follow advances in technology closely and continually try to push its capabilities to their limits; here the word of caution is that when you work at the cutting edge of technology, it is important to stay *behind* the blade. Remember that fundamental change is slow and somewhat unpredictable; it takes time for new ideas to propagate, sink-in, and gain acceptance.

We must be continuously watchful to ensure that we match our adopted information and communications technologies to the needs of our clients. The keys to success are sound strategy, an understanding of the big picture, adaptability, and a long view of the future. Even our entrepreneurial hero, Aldus Manutius, seems to have understood this. The logo of the Aldine Press featured a dolphin (for speed?) wrapped around an anchor (for stability?), and Aldus' motto was "make haste ... slowly." Think about that, and may the future be your friend.

ELECTRONIC JOURNALS: NEW HEADACHES AND OLD

Judy C. Holoviak

American Geophysical Union
2000 Florida Ave., N.W.
Washington, D.C. 20009

Abstract – Planning an electronic journal has all the headaches of launching any print product and many more. Who will be the audience? How can authors be convinced to submit? What is the financial outlook? Are there any special characteristics of the production process to be considered? These are typical questions that society and commercial publishers have learned to address. The new media brings new questions. How do we go about reviewing material that cannot be printed? Must we limit reviewers to those who have the right software interface? For the first time we are forced to face the creation of a technological elitism in the dissemination of scientific journal literature. Will we be able to recover the cost of publication from subscription fees, particularly when the culture of users to date has been to consider electronic information as free? The advent of the serious electronic journal is further inhibited by other less evident obstacles. U.S. law is unclear in the area of copyright, and the issues of copyright protection become much more complex as we move across national boundaries. There are significant financial risks in producing a new electronic journal. It is little wonder that progress toward fully electronic publishing is moving more slowly than technology would seem to indicate is possible.

These would be challenging times without the electronic revolution. The volume of material being produced by today's researchers is beginning to outstrip our ability to cope with it. Individuals can no longer keep up with the breadth of the Earth sciences. To keep up, many find they must focus their interests more and more narrowly. Their shelves are overstocked. They are no longer able or willing to purchase personal subscriptions to the journals they would like to have at their finger tips. I do not need to remind the Geoscience Information Society (GIS) audience about the problems of shrinking library budgets and increasing volume and variety of material – or the cancellations that are occurring.

There is more material to be published and fewer subscribers to help pay the bills. As publishers of scientific journals, we will need to keep the current system going while we are adopting the new technologies. This means that we will be incurring added costs at a time when our markets are shrinking. Individuals have come to think that information they receive electronically should be free – at least the online information should be free. At a minimum there is an expectation that electronic publishing will reduce costs dramatically. Experimentation is risky and there is not much evidence about which of the different electronic modes is likely to be most successful. This is hardly the financial climate one would choose for introducing major, untried changes. But we could have little choice.

Particularly for those of us involved in societies, early – maybe even premature – change could be mandated by our boards of directors. The researchers on these boards are becoming accustomed to almost instantaneous access to information from anywhere in the world at their own desks. They have seen the power of electronic communication and will expect this power to be used for journal literature as well.

So the expectations of our boards might be the first headache we face. But there are many more.

It is easy to become enthusiastic about the wonders of the new media. But there are many questions to resolve before serious journals will be comfortable with electronics as the primary means of publication.

For starters, we will need to decide which of the new media we will select as our delivery vehicle. Will our journal be delivered via CD-ROM or will it be an online publication? Because I am particularly interested in online delivery, I will concentrate on that mode in this article. But many of my remarks will also be valid for CD.

Headache number 2 is the lack of a standard interface for delivery of our online journal. Many players are experimenting with different kinds of software for selection and display. The Online Computer Library Center (OCLC), which has traditionally been a service organization for the library world, is moving aggressively to be the big player in providing the hardware and software for publishers of online journals. They with American Association for the Advancement of Science (AAAS) created *Online Clinical Trials*, the prototype journal for OCLC's system. The first display system was excruciatingly slow and turned sophisticated users off. I understand that the current version, which is more costly for them to operate, is much better. Elsevier has just announced their plans to offer at least one journal starting in 1995 via OCLC's *Guidon* system and may have other journals in the wings. Two of the physics societies are developing experiments with OCLC now.

AT&T is offering First Page. This software is definitely user friendly but does not seem to be capturing the scientific journal market, although some scientific societies have been or are still discussing experiments with this software.

Mosaic is taking the online world by storm, but it has major drawbacks for scientific material. Although Mosaic can wow us with moving images and sound, HTML is very limited in the fonts that it can support. It cannot handle Greek characters or most math symbols. Every occurrence of Greek and all but the simplest equations must be treated as graphical elements. Unless these are handled well, the display can look like it is spotted with tiny halftones, a state that certainly impairs readability. But more serious than the poor visual presentation is the fact that we cannot search the content of the graphical representations of the math.

Bellcore with the American Chemical Society and the Cornell Science Library has engaged in extensive testing of user interfaces. We may see some interesting things come out of that collaboration.

For publishers, selecting the right interface will be tricky. Changing it will not be as simple as moving a journal to a new supplier of printing services, which can be a significant headache. Today, we need to select an interface that handles our material well and that we think will be attractive to our readers. We need to consider the cost of using the interface. How do those costs fit our financial projections? Are we comfortable with the policies and practices of the vendor? We may be together for a long time, if we are lucky. Additionally, we need to remember that we are gambling on who will be a winner in the sweepstakes for the most successful user interface.

How our journal will appear to the user will be determined in large measure by the user interface we select. But we will want to consider ways to make the screen display as readable as possible Will the display be the same as the print image we provide for onsite printing? Most users will probably not want to do all of their reading on screen. They will browse or search online but will probably print the article for ease of reading and study, presuming the journal publishes material that can be printed. If, however, the electronic journal is simply a screen representation of what can be conveyed in the flat surface of the printed page, publishers are not exploiting the technology.

If we take the easy road and limit ourselves to material that can be printed on paper, we will simplify our life. If not, we will need to find a new kind of reviewer. Editors will not only need to consider matching scientific expertise of the reviewer to the subject of the paper, but also need to be sure that the reviewer's hardware and software can handle the submission. Editors may find their pool of reviewers restricted by technological compatibility. For the first time in the sciences we will be forced to face the possibility of the creation of technological elitism in the dissemination of scientific journal literature. There will be interested scientists who simply will not have the technological wherewithal to access the new electronic journals.

There are smaller headaches as well for editors. They will need to be sure that the material sent out for review is virus-free. Editors cannot be sure where authors have picked up all the material that was included in the electronic submission. Can you imagine the havoc that a disgruntled author might cause an editor by planting a virus that the editor will spread? That certainly is not a way for editors to ingratiate themselves to reviewers.

Once the article is published, the job is not over. Not in the online journal. The publisher must be sure that files have been prepared and properly archived for the future. This may be something that we will contract to a vendor, but we will want to confirm that the terms of the agreement to provide this service are being followed. When we are ready, or are forced, to migrate the journal to a new platform, we cannot afford to rebuild the back files of material.

I see the publisher of the online journal taking on the responsibility for providing the authoritative source for the future. Copying and redistributing online material is extraordinarily easy. In this process, journal articles could be inadvertently or deliberately altered. How can the reader be sure that the version he or she has received is the same as what was published? Some organization must take the responsibility for assuring the integrity of the primary source. Many scientific societies expect to assume this responsibility as part of their obligation to their authors and readers.

We are beginning to hear more about electronic fingerprinting of material. This technique would help trace the source of the material being distributed. The technique is being considered as a way of assuring that copyright can be protected in the online environment. It may also have a role to play in the type of authentication I am considering.

There are a few other technological issues I would like to explore, but I am going to hold those off for a bit. Since I have mentioned copyright, I would like to review a few of those headaches.

Copyright had become a pretty ho-hum topic after the flurry of debates sparked by the omnibus revision to the copyright law in 1976. Today we hear many proponents of abandoning copyright. They say it has no place in the electronic world either because it inhibits the freedom of electronic communication or because it cannot be protected in the electronic world and therefore is relevant. Those in the latter camp may be supported by the contingent who fought the privatization of the Internet; these folks put forth the argument that "the demise of a free Internet will violate their first amendment rights."

Despite all the rhetoric, the recent report from the government's Task Force on the National Information Infrastructure (NII) has come down firmly and unequivocally on the side of protection of intellectual property in the online world. They have said that the success of the information superhighway will be driven by the content moving over it. In short, their report finds "the potential of the NII will not be realized if the content is not protected effectively" (Working Group on Intellectual Property Rights, 1994).

The Task Force has found that the Congress was quite forward-looking when it drafted the current copyright law. Congress tried to anticipate new technologies and to be certain that the law would not be easily outmoded by what the future had in store. After all, Congress was drafting the law at the time when the photocopying machine had raised havoc with the definitions under the prior law. In fact the Supreme Court in its decision on Williams and Wilkins vs the National Library of Medicine had urged Congress to clarify the law. Congress did well.

In its recent report, the NII Task Force noted that most of the principles in the current copyright law hold very well in the online world so long as we understand some of the implications. For example, there is a doctrine of first sale, which states that "The owner of a particular copy of a copyrighted work may distribute it freely, but may not reproduce it." This means that if I purchase a copyrighted publication, I have the right to dispose of it as I see fit so long as I do not make a copy. I do not need the permission of the owner of the copyright. However, in the digital world I must remember that I do not have the right to transmit a copyrighted document that I have acquired legally unless I purge my digital copy; otherwise I have indeed created a copy and I do not have that right. It is easy to forget to erase material and, if I forget, I have broken the law.

A particular worry for the providers of serious online journals is the ease with which material can be retransmitted to a few or many others. Under the current copyright law I do not have the right to do this. It is tantamount to reproduction and I do not have that right unless the copyright owner has explicitly granted it to me.

Some are suggesting that we need to provide a new definition of "publish" for the digital world. They are urging that Congress revise the copyright act by adding "transmission" as one of the methods of distribution that can constitute publication. Whereas those suggesting this so-called clarification may be well-intended, in fact such a change could be quite detrimental to the scientific process. We would not want to see authors' current practice of sending their papers to a few colleagues for comment thwarted. This type of distribution should not constitute prior publication. At the same time, there is concern about the role of the systematic preprint services that broadly distribute material prior to formal publication. But that is a can of worms far beyond the scope of this article.

At the American Geophysical Union (AGU), we think that the reproduction right under the existing law can be used to provide adequate protection of copyrighted works in the network environment at least with current technologies. We would not favor the proposed revision to change the definition of publication in the law.

I do not want to leave the reader with the notion that copyright is no problem for the electronic journal. We need to be certain that there is a penalty for those who would remove or destroy the copyright notice on digital documents. A change in the law to provide such a penalty should be supported. This change would be for the protection of the creator of the document, for the protection of the users of the document, and for the protection of any revenue that the copyright owner hopes to acquire.

The major headaches come into play when we consider the myriad of copyright laws around the world. What may work under U.S. law may be quite impossible in the U.K. We also need to consider the fact that our electronic journal will be accessible to folks in countries that have no copyright law and therefore provide no protection to the material. International copyright considerations will be far more important in the electronic environment than they have been in the print one.

For this reason it is very likely that publishers will consider license agreements for their electronic journals. In a license agreement the rights can be clearly defined for all. And yes, it is possible to take away users' rights or give away owners' rights in these agreements.

Nonetheless such contracts are easier to understand and easier to defend, should it become necessary.

The publishers of online journals will deal with new types of copyright permission. It is unlikely that each publisher will create the software needed to produce, store, and distribute his or her online journals. Nor will all the software that we want for our journal distribution be in the public domain. Publishers will be forced to enter into license agreements with software vendors both for their own use and possibly for the use by their subscribers. These licenses will add to the cost of publishing.

Since I have mentioned costs, I think we should leave copyright headaches behind and turn to some of the financial ones.

Many proponents of electronic publication expect tremendous cost savings and therefore major decreases in their subscription fees, particularly in the online world. After all is not the Internet free? Why should there be any costs borne by the end users?

The arguments that I have heard put forward in support of huge cost savings are to my mind rather naive and unrealistic. Most publishing costs will not go away and new ones will be added.

In journal publication about 80% of the current cost arises from editorial selection, copy editing, and preparing material for printing. These costs will remain when we publish in electronic format. We will still have the editorial selection, copy editing, and preparing material for distribution.

New costs will be added for improved indexing and other tools needed for retrieval. If we are to take maximum advantage of SGML, we will need to spend time and money to be sure that the material has been properly tagged. The tagging that is provided as a byproduct of composition is not sufficient.

And what about customer support? That is not an item publishers worry about in the print environment. I read recently that the Johns Hopkins University Press has added systems manager and technical support specialist to their list of job titles. Some of this type of work may be provided by a vendor, but the publisher and ultimately the subscriber will pay.

Depending on the vendor selected for distribution and the complexity of the material in the journal, there may be significant capital expenditures in shifting to online publication. For example, OCLC did not have the fonts needed for extensive mathematics. I do not know the details of the financial arrangements, but I understand that it took about $300,000 to produce the fonts for one physics journal. Someone, sometime must pay for that investment.

We must plan and pay for systems that will keep the digital files in good condition. (We all know horror stories about magnetic tapes that have deteriorated beyond use.) We must also consider the expense of shifting materials to new platforms. Serious journal publishers must build this expense into their financial models. We must collect the money up front for taking care of the digital material. We will be paying for an active custodial process, one that checks on the quality of the storage medium and regularly refreshes it. This process may also require moving to new forms of storage as old systems become obsolete. I believe that subscription fees will contain an element of cost for this "perpetual care"; otherwise we cannot be sure that we can serve tomorrow's subscribers.

For publishers who choose to provide both print and online versions of the same journal, there will be minimal savings on the print side (possibly less paper and postage) but increased costs for the dual electronic production.

It is likely that some payments will shift from up-front subscriptions to pay-for-use. This is a radical change in the economic picture. Subscribers expecting to see savings from such a shift are overlooking the fact that they will have to pay something just to have material available when they want to use it.

Developing pricing models will not necessarily be easy. In the institutional market, what constitutes a subscriber? Will we permit log-ins from any computer or will we restrict them to one computer, one building, one campus? Is the University of California system one subscriber? Must we consider the number of simultaneous log-ins in our pricing formula?

Subscribers are going to want to know what they have paid for when they subscribe to an online journal. Will the subscriber have the right to print unlimited copies? What about access in the future? Does the subscription fee give access only for the year in which it was paid? What about back issue or back year access? The publisher will need to consider these questions in developing the pricing model and will need to explain the conditions very carefully and very clearly to the subscriber.

The financial risks increase as we move into new territory. The up-front costs of product design and testing may be quite high. It may be difficult to get good marketing information without trying some experiments. Many of our customers have been conditioned to expect online information to be free. The most exciting delivery system today in the online world, Mosaic, does not yet have a reliable mechanism for charging.

The financial side of the online journal presents headaches for all of us—just as it does in the print world.

By this list of technological, legal, financial and other headaches, I seem to have created quite a gloomy image of online publishing at least from the perspective of a publisher. You might ask why in the world would anyone go into it. The answer is easy.

We have an opportunity to provide real added value for the authors and the readers, especially if we are willing to stretch our imaginations a bit.

Let me mention some of the easy things first. We can reach all subscribers around the world at the same time. For AGU, more than half of our library subscriptions are overseas. We are competing with European commercial journals that are produced on the continent and are distributed much faster than the U.S. journals. If we add the cost of air shipment to get the journals to the reader faster, we will lose subscriptions. As speed of publication becomes more and more important, there will be competitive advantages to shortening the delivery time.

Storage is another of the major reasons for declining subscriptions. Individuals who subscribe to journals are running out of shelf space, at least they are if they take the *Journal of Geophysical Research*. Libraries cannot continue to build warehouses for their holdings. It seems to me that if libraries can cut the costs of processing and storing issues, they will have more to spend on materials. Even if subscription prices remain the same, I believe that the total cost to the library should be less.

Some of us will provide online journals because it is expected. Our authors and readers see this as the wave of the future.

The excitement of online publishing is, however, the chance to provide added capabilities for the user. The best online journals will be much more than a visual representation of the printed page on a computer screen. The truly exciting journals will provide "hot material." They will give the reader access to the data underlying the figures in a way that he or she can actually use them. Mathematical expressions will be manipulatable. Readers will be able to try their own data in the models or pick up the data from a table and use those data in other models. There will be links to external databases. These journals will publish three-dimensional and moving representations of data. In short, the online journal will publish material that cannot be printed.

Authors will no longer be forced to reduce the value of the information they include in their articles. The power of the computer they use in analyzing and displaying their research results will be available to the reader.

Are there such journals today? Not that I know of. Will there be soon? I believe so. AGU is currently working with the American Meteorological Society and several other scientific societies to create just such a journal. We know that there are authors waiting to publish this kind of material. There are research results that simply cannot become part of the record because there is no vehicle for them.

- Do we think this journal is going to be easy to produce? Not really.
- Do we think that it will be costly to develop? Yes.
- Are we fully aware of and prepared for the headaches? Probably not, but we are trying to anticipate them.
- Are we excited about the prospect of serving Earth scientists who are no longer satisfied with material presented in Flatland? You bet we are.

REFERENCES

DeLoughry, T.J., 1994, "Fair use" for electronic age: Debate over copyright law heats up: The Chronicle of Higher Education v. 41, no. 5, p. A30.

Lewis, S., 1994, From Earth to ether: One publisher's reincarnation: NASIG 9th Annual Conference, The NASIG Newsletter, v. 9, no. 4, p. 21.

Ricart, G., 1994, Free the Internet!: Computers in Physics, v. 8, p. 508-510.

Riordan, T., July 7, 1994, Writing copyright law for an information age, New York Times, p. C1.

Risher, C.A., and Gasaway, L.N., 1994, The great copyright debate: Library Journal, v. 119, no. 15, p. 34-37.

Woo, J., 1994, Government paper on copyrights in cyperspace vexes some firms: The Wall Street Journal, Sept. 2, 1994, p. B3.

Working Group on Intellectual Property Rights, 1994, Preliminary Draft of the Report of the Working Group on Intellectual Property Rights, chaired by Assistant Secretary of Commerce and Commissioner of Patents and Trademarks Bruce A. Lehman, Executive Summary: July 1994, p. 2.

THE EFFECTS OF TECHNOLOGY ON THE
INFORMATION-SEEKING BEHAVIOR OF SCIENTISTS

Julie Hallmark

Graduate School of Library and Information Science
The University of Texas at Austin
Austin, Texas 78712-1276

Abstract – This paper examines scientific communication in all disciplines with an emphasis on the Internet. An overview of the wide variety of resources offered on the Internet for one discipline (chemistry) is followed by descriptions of general applications, including collaboratories, electronic publication, online conferences, document delivery, and data and image access and retrieval. The author suggests some problems and issues which have arisen in connection with scientists' use of technology and proposes factors which may encourage (or discourage) the use of new technological options. These factors include differences among disciplines, specific individuals in the organization who lead the way, the commitment of the organization to new technology, cost and ease of use, geographical location, and corporate policy.

INTRODUCTION

Technology is changing the face of scientific communication Using fax, online and CD-ROM services, and e-mail and other Internet resources scientists can rapidly access and retrieve files of data and text, communicate instantaneously with colleagues, download software, search databases, participate in conferences and work sessions, post queries on bulletin boards which prompt hundreds of replies, and access electronic journals and preprints. Commenting on the Internet, astrophysicist Larry Smarr stated:

It's the most fundamental shift since Gutenberg. The Internet is basically a space and time destroyer. It shrinks distance and time to zero. It's as if all the world's scientists were in one room, available at one computer. Needless to say this is having a profound impact on the way science is done (Broad, 1993).

Technology has also revolutionized the access and retrieval of published scientific journal articles, still the most critical mode of scientific communication. New options for document delivery have become crucial services in times of shrinking journal collections onsite. As the popular predictions of "libraries without walls" and "access rather than holdings" become realities, scientists must continue to have efficient and cost-effective journal article retrieval.

Illustrating new patterns and challenges, this paper examines scientific communication in all disciplines with an emphasis on Internet applications. Selected examples show the broad range of possibilities now available to scientists.

APPLICATIONS

The Internet has had an overwhelming influence on scientific communication (Markoff, 1992). Robert Snyder, director of the Institute for Ceramic Superconductivity at Alfred University in New York points out that "My entire operation has been transformed by Internet." He keeps in constant touch with graduate students training overseas and with those at Alfred when he goes to conferences. He also uses the Internet to co-author papers with colleagues in Moscow and Eastern Europe who were formerly difficult to communicate with. "In the past, using land mail, I had a couple of colleagues that I collaborated with internationally, but we couldn't do much. Now I have about 25 active collaborations going on around the world because of Internet. I can't live without this thing (Broad, 1993).

To illustrate the wide variety of Internet resources available in a single discipline, Mosaic was used to uncover the following examples in chemistry. They represent only a part of the total resources for that field:

Listservs Medicinal Chemistry, Computational
  Chemistry, High Resolution IR
  Spectroscopy, Radiochemistry
Newsgroups - Organometallics, Polymers,
  Agriculture, Microscopic Techniques
Preprints - Chemical Physics, Nuclear Theory
Journal Articles and Abstracts - Chemical
  Reviews, Journal of the American
  Chemical Society, Journal of Porous
  Materials

Text Databases - Agricola, ENZYME,
Quantum Chemistry Acronyms
Numeric and other Data - Genome, Nucleic
Acids, Hazard Data, Materials, Periodic
Table
Newsletters - Catalyst, Computational
Chemistry, Computers in Chemistry
Conferences - CHEMCONF, Computational
Chemistry, Chemometrics
Indexes and Tables of Contents - Biochemistry
Journal, Chemistry Textbooks in Print,
International Journal of Quantum
Chemistry, Journal of Biological
Chemistry, Journal of Computational
Chemistry, Springer Journals
Reference and Document Delivery - Patents
Copy Service (Derwent), Chemistry
Information (San Diego State University),
Clearinghouse for Chemical Information
Instructional Materials

Collaboratories

The notion of a collaboratory embodies many of the activities discussed in the previous sections and more. First proposed by Wulf in 1989 a collaboratory (a blend of the words collaboration and laboratory) was defined as:

> a center without walls, in which the nation's researchers can perform their research without regard to physical location--interacting with colleagues, accessing instrumentation, sharing data and computational resources, and accessing information in digital libraries (Wulf, 1993).

A report from the National Research Council published last year expands Wulf's idea and suggests attributes and tools desirable for building collaboratories in oceanography, space physics, and molecular biology. Drawing on existing resources in those fields and pointing out problems of hardware, software, and personal and social factors, the report envisions virtual laboratories using networked facilities which "render irrelevant the actual location of equipment and instrumentation" (National Research Council, 1993).

Wulf points out that the bottleneck in building collaboratories is not the physical infrastructure of networked computers augmented by instrumentation interfaced to the network, but the software required for access and use of huge scientific databases such as the genome project or global seismic data. These databases have different attributes and requirements (such as simultaneous, multi-dimensional access) than allowed by typical database management systems. What he terms "data fusion" results from the existence of large amounts of online data which must allow for the analysis of phenomena observed by multiple sensors in multiple locations. The data must not only be accessed and manipulated but also understood. Sharing the results of remote instrumentation such as deep space probes presents another set of challenges (Wulf, 1993).

Electronic Publication

Textbooks and reference works are becoming more widely available on the Internet, as well as publication in a combination of media. For example, Jon Rathblatt, an assistant professor of biological sciences at Dartmouth College is coediting a book on genes and proteins of the secretory pathway to be published in hardcopy, but updates will be posted on the Internet (Hoke, 1994a).

Preprint bulletin boards in physics are perhaps the most currently successful and heavily used electronic publications. Created by Paul Ginsparg at Los Alamos, bulletin boards representing such subdisciplines as nuclear physics, astrophysics, condensed matter physics, high energy physics--lattice, and high energy physics--theory allow physicists to submit preprints, scan preprints already on the system, read abstracts that sound interesting, and request that an entire preprint be transmitted which takes a second or so and is free of charge. The Los Alamos National Laboratory Physics Information Service also provides links to similar servers in other locations (Taubes, 1993a).

Read by scientists in some 50 countries the service has revolutionized the old model of the invisible college where one received preprints by virtue of being on a very select list, "introduced" to the group by one's supervising professor or mentor. The popularity of the preprint bulletin boards was dramatically demonstrated last year when Ginsparg shut down the system for a few days due to lack of support and funding. The closure incited physicists all over the world to petition the Department of Energy and the National Science Foundation for funds; the project also secured official funding from Los Alamos (Taubes, 1993b).

Physicists, of course, have had a preprint culture for years, citing the advantage that preprints are available months (or years?) earlier than peer-reviewed journal articles. Andy Strominger, a physicist from the University of California, Santa Barbara says:

> In the dark ages people used to publish in journals. Then came this preprint system.

Copying got cheaper and everybody started mailing out copies at the same time they submitted them to journals. Now everybody sends their papers to Ginsparg's bulletin boards (Taubes, 1993).

Online Conferences

In 1993, the American Chemical Society's Division of Chemical Education sponsored the first experimental online electronic conference entitled "Applications of Technology in Teaching Chemistry," conducted entirely by means of network services. Some 450 educators from 33 countries participated—impressive statistics for a specialized chemical education meeting consisting of only 15 papers (O'Haver 1994). Scientific papers and conferee questions and responses were transmitted by e-mail. Access to papers, graphics, instructions, software, and discussion transcripts from the conference was through direct file transfer from the University of Maryland computer system.

The conference began in June when registrants received abstracts of the scientific papers and instructions on how to receive the electronic versions of the full text and graphics. The "meeting" consisted of three three-week sessions of five papers each. The first week everyone read papers; the second and third weeks consisted of extended two-day discussions held about each paper; a general one-week discussion period closed the conference. All dialog was archived by the University of Maryland.

Participants concluded that the author-listener format allowed a deeper and more detailed examination of the papers and more productive interactions than at a traditional conference. "No sooner did I ask a question than an answer appeared on the screen." Discussion was one-on-one and backup materials were available. Attendees had no feeling of being rushed, and people who had little money to attend conferences could participate. One deaf participant especially appreciated the format (O'Haver, 1994).

Negatives included the huge e-mail burden, the submission at times of long-winded irrelevant messages (a presenter might find up to 40 messages waiting to be answered), and the complexity of accessing graphics. Some participants missed perquisites such as expense accounts, good restaurants, and hotels.

Borman (1993) has suggested that a meeting could be conducted in two parallel formats—electronic and regular. "Those who are able to attend can go; those who aren't can still read papers, ask questions, and even communicate with the authors personally."

Document Delivery

Over the last two decades prices of scientific journals have risen by an average of 13.5% per year. Today an annual subscription costs the library $1000 to $1500 or more, and in one year recently, members of the Association for Research Libraries cancelled a total of 60,000 serials. The financial crunch extends to other countries too, of course; academic librarians report that visiting foreign scientists devote a great deal of their time in the U.S. photocopying articles in the library. Jackson (1993) has pointed out several new models of document delivery which have been implemented in response to the serials crisis:

- use of commercial document delivery services or free services such as Ariel through the local library
- a shift of responsibility to the user by mounting table of contents services, such as Current Contents, as part of their online catalogs of campuswide information systems. Users may directly access such services as OCLC's First Search or RLIN's CitaDel
- a combination of traditional interlibrary loan and other modes of delivery.

Ariel, perhaps the most popular of the new approaches in academic libraries, was pioneered by the Research Libraries Group (RLG) and supports digital transmission of documents over the Internet. A staff member at the supplying library scans articles directly from the bound volume, and a three-letter code directs transmission to the requesting library. The articles are compressed to less than 10% of their original size for rapid transmission in digital form, conveyed over the Internet, and then printed out on a laser printer at the receiving library, requiring no human intervention. The high-quality, high-resolution copy is a striking advantage of Ariel, especially valuable for scientific graphs, illustrations, and photographs. Less than 2% of document transmissions need to be repeated (Bennett and Palmer, 1994).

Scientists still lament the loss of subscriptions in paper, but are more resigned to the situation than they were a few years ago. Cost remains a problem for many academic scientists in using some of the new document delivery options on their campuses. One librarian reported that her biology clientele balks at paying the $15-25 required for their rapid document delivery service. UnCover fees, for example, are currently $16.50 per article, which seems unacceptedly high to many users (Whittaker and Malamud, 1994).

Many scientists hope for extensive full-text transmission of journal articles via the Internet in the future. The frequent assumption that such retrieval will be free is simplistic; publishers do expect to continue being reimbursed for their journals no matter how they are accessed. Other unresolved issues include: the best means for accomplishing satisfactory peer review; the need for permanent archiving of science; and assuring access to users who do not have electronic capabilities.

Data and Images

Thousands of scientific data projects are being implemented on the Internet through government agencies, universities, research laboratories, institutes, associations, and other organizations. Offering free access, no limitations on size, and capabilities of instant revision, such databases offer great promise and convenience to researchers. A few representative examples serve as illustrations.

Last December in *Nature* Daniel Cohen announced completion of the first draft map of the human genome (Cohen and others, 1993). Covering 20,750 overlapping segments of genome from ten different humans, the map included 2000 markers that serve as signposts for specific regions of the genome. "The sheer size of the project made [Internet] the only option for disseminating the information," Cohen pointed out. "If printed on paper the map "would create a pile several hundred meters high." On the day the paper was published 64 requests for map information were received (Nowak, 1993). The Human Genome Project, with the goal of identifying all of the estimated 100,000 human genes by the year 2005, will continue to produce large amounts of data, much of which will not have equivalents in the printed literature. These enormous data collections will be accessible by a variety of options (Primich, 1994).

The National Library of Medicine is creating a national standard for human images. The Visible Human Project will provide a road map for developmental biologists--scientists studying gene expression which will describe their findings using Cartesian coordinates on a standard anatomy, rather than using more general verbal descriptions of a location within an anatomical structure. The goal is to link transparently the print library of functional-physiological knowledge with the image library of structural-anatomical knowledge into one unified resource of medical information (Miller, 1994).

.The Comprehensive Oil & Gas Information Source (COGIS) contains virtually all of the oil and gas data published for the last decade by the Energy Information Administration (EIA) in: *Petroleum Supply Monthly, Petroleum Marketing Monthly, Natural Gas Monthly,* and *Weekly Petroleum Status Report.* Developed by the EIA in cooperation with the Department of Commerce and available through the Commerce Department's Economic Bulletin Board, COGIS covers oil reserves, production, refining, products, distribution, storage, transportation, imports and exports, consumption, and prices (Energy Information Administration, 1993).

PROBLEMS AND ISSUES

One may ask the critical question: To what extent are the vast scientific resources on the Internet actually being used by scientists? One recent research study estimated that the routine use of computers in support of research and production in a chemistry laboratory or office, other than for word processing, spreadsheets, and literature searching, is less than 25% of the potential users (Heller, 1994). On one large university campus in the U.S. only 20 of the 60 chemists on the faculty had e-mail addresses in 1994, and many of these had been "automatically assigned" in connection with a project and had never been used. The American Chemical Society Computers in Chemistry Division (COMP) distributes its newsletter via e-mail as well as hard copy; in 1994 only about one-quarter of COMP members received the electronic version. And a biology librarian estimates that no more than one-third of the scientists on her faculty have e-mail accounts. Thus, many potential users do not take advantage of the Internet, for one reason or another.

Other problems include:

- Data overload and "psychological burnout." One scientist who received more than 500 messages daily stopped using the Internet altogether (Broad, 1993).
- Difficulty using e-mail to initiate research projects. Some scientists say that when they are planning their work, they need and rely on more interactive and expressive forms of communication. With e-mail, they also report that they make "slower progress than when they bump into each other in the halls every day" (Hoke, 1994b)
- Resources which sound promising, but turn out to be trivial or offer little of value. "Elsevier tables of contents--big deal!" says one biology librarian. Or they may appear redundant; one chemist wondered why he would want to access the Periodic Table online when it sits on his desk in a handbook.
- Elitism on the Internet. People who are doing the really great work in a field are not on the bulletin

boards, but corresponding via personal e-mail. "We still have the haves and have nots," a chemist commented.

CONCLUSION

As the present study progressed, it was interesting to speculate upon the factors which encouraged (or discouraged) the use of new technological options. Thus, in conclusion, let us propose some possibilities:

- The discipline itself and the degree of desperation or need to access the newest research as quickly as possible. Rapid developments in a field are the most powerful motivation to learn new tools of access and retrieval. A colleague only somewhat facetiously correlates the speed of objects studied by the discipline to perceived need for rapid access. So with objects ranging from high energy particles to glaciers, one can draw conclusions appropriately. Chemistry and biology lie closer to high energy physics at one end and mathematics closer to geology at the other.
- The presence in the department or lab of a gatekeeper or information junkie who takes the initiative for acquiring and learning a new tool and teaching its use to others.
- The commitment of the organization to new technology. None of the faculty in one college of Pharmacy were on the Internet because the local area network was only internal to the college and the dean saw no reason to spend more money on communications hardware and software.
- Cost and ease of use. User-friendly software such as Mosaic which simplifies access to the Internet has revolutionized network use. Developed by the National Center for Supercomputer Applications at the University of Illinois, Urbana-Champaign, Mosaic allows a graphical, intuitive approach. Browsing allows users to move through online resources linked by hypertext (Krumenaker, 1994).
- Geographical location. Scientists in the Peoples' Republic of China, for example, have felt cut off from electronic options. That country is just now acquiring an adequate telecommunications infrastructure. Australian scientists, on the other hand, are very heavy users of the Internet, relatively. For example, Dr. Brendon McKay, a computer scientist at the Australian National University once received an electronic query about his work from Dr. Stanislaw Radziszowsky, a mathematician at Rochester, since then they have exchanged over 1000 messages.... "While one works the other sleeps" (Broad, 1993).
- Corporate fears of invasion of their computer resources by the outside world, whether by viruses or by competitors. To many (who pride themselves on their protective "firewalls"), allowing access to the Internet appears risky.

There is no doubt that extensive scientific resources are now accessible through the Internet and other new technological applications. Some important questions remain, however, concerning: integrity of the scientific record, access to everyone regardless of ability to pay (formerly provided by libraries to a great extent), appropriate means of reimbursement for the commercial sector, user training, and the future of the scientific journal and other traditional communication media. It will be fascinating to observe how the solutions of these issues evolve.

REFERENCES

Bennett, V., and Palmer, E., 1994, Electronic document delivery using the Internet: Bulletin of the Medical Library Association, v. 82, p. 163-167.

Borman, S., 1993, Benefits of electronic conferences cited by on-line meeting 'attendees': Chemical and Engineering News, v. 71, no. 41, p. 26.

Broad, W.J., 1993, Doing science on the Internet: a long way from Gutenberg: The New York Times, May 18, 1993, p. C1.

Cohen, D., Chumakov, I., and Weissenbach, J., 1993, A first-generation physical map of the human genome: Nature, v. 366, p. 698-701.

Energy Information Administration, 1993, Comprehensive Oil & Gas Information Source (COGIS) [pamphlet].

Heller, S.R., 1994, Gazing into the future of chemical information activities: in Collier, H., ed., Further Advances in Chemical Information: London, Royal Society of Chemistry, p. 16-34.

Hoke, F., 1994, Scientists predict Internet will revolutionize research: The Scientist, v. 8, no. 9, p. 1.

Hoke, F., 1994, Publication by Internet: The Scientist, v. 8, no. 9, p. 8.

Jackson, M.E., 1993, Integrating ILL with document delivery: five models: Wilson Library Bulletin, v. 68, p. 76-78.

Krumenaker, L., 1994, New software and services ease access to the Internet: The Scientist, v. 8, no. 9, p. 17-19.

Markoff, J., 1992, A network of networks that keeps scientists plugged in: The New York Times, January

1, 1992, p. 49.

Miller, J.A., 1994, Anatomy via the Internet: Bioscience, v. 44, p. 397.

National Research Council, 1993, National collaboratories: applying information technology for scientific research: Washington, D.C., National Academy Press, 105 p.

Nowak, R., 1993, Draft genome map debuts on Internet: Science, v. 262, p. 1967.

O'Haver, T.C., 1994, CHEMCONF: an experiment in international on-line conferencing: Abstracts of Papers of the American Chemical Society, v. 207, 40-CINF.

Primich, T., 1994, Genetics databases available via the Internet: Abstracts of Papers of the American Chemical Society, v. 207, 41-CINF.

Taubes, G., 1993a, Publication by electronic mail takes physics by storm: Science, v. 259, p. 1246-1248.

Taubes, G., 1993b, E-mail withdrawal prompts spasm: Science, v. 262, p. 173-174.

Whittaker, M., and Malamud, J., 1994, UnCover: the article access solution: Bulletin of the Medical Library Association, v. 82, p. 181-182.

Wulf, W.A., 1993, The collaboratory opportunity: Science, v. 261, p. 854-855.

PART II

TECHNICAL SESSION:

GEOSCIENCE INFORMATION

IDENTIFYING CORE GEOLOGIC RESEARCH JOURNALS: A MODEL FOR INTERLIBRARY COOPERATIVE COLLECTION DEVELOPMENT

Louise S. Zipp

Collection Development Department
204 Parks Library
Iowa State University
Ames, Iowa 50011-2140

Abstract — While scientific journal prices continue to rise, the level of funding support for these materials in library collections declines. Many libraries undertake circulation and on-site use studies to identify journals for cancellation. The results tend not to be comparable because each study is designed to fit unique circumstances of information delivery. Subsequent decisions do not necessarily reflect the research needs of the primary user group. As research journal subscriptions are canceled, cooperative collection development of these titles becomes necessary to restore depth and breadth of collections. This study develops a means to identify core research journal titles used by the primary geologic user communities at the three public universities in Iowa. For a three-year period, several products of research journal usage were analyzed: citations from faculty publications and grant proposals, citations from theses and dissertations, and citations from student publications. Preliminary results showed different local collection needs. The core lists were used to identify gaps and redundancies in their respective collections. Unique titles on grant proposals pointed to increasing interdisciplinarity in geoscience research. Although a low response rate compromised the validity of faculty data, the assumptions, methods, and preliminary results were evaluated for further use.

INTRODUCTION

At the 1993 Geoscience Information Society Technical Session, Noga, Derksen, and Haner presented a paper entitled "Characteristics of geoscience serial use by faculty and students". Their important study showed the results of different measures of serial use tested at Stanford University and at the University of California at Los Angeles. They concluded that titles cited in theses, dissertation, and faculty publications represented only one-third of the total number of titles used during the study. Those titles, however, included some of the more expensive, esoteric research serials in the collections.

I was prompted to ask the question: Can citation analysis be used to assist cooperative collection development by defining the research journal collections and needs at different university libraries? Specifically, I wanted to develop a method that would enhance the modest level of cooperation now in place among the research libraries at the three publicly-funded universities in Iowa: Iowa State University (ISU), the University of Iowa (UI), and the University of Northern Iowa (UNI). All three institutions are governed by the State Board of Regents. Both UI and ISU support doctoral programs in geology; UNI offers a Master's degree.

In addition to my primary question, I also sought to answer the following questions:

- Is there an identifiable Regents' library collection of research journals in geology?
- Are there distinctly different collections and needs among the three universities?
- Do grant proposal citations show the coming trends in research at the universities?
- Can library collections presently support those trends?
- For titles identified as potential candidates for cooperative collection development activities, are there characteristics about them or their use which complicate sharing?

CHARACTERISTICS AND DESIRED RESULTS OF A SUCCESSFUL METHODOLOGY

- Adaptable to different libraries, particularly where decentralized collections exist at some sites.
- There is access to full population of citation data or clear knowledge of gaps in coverage of research interests.
- Principal investigator need not have in-depth subject expertise.
- Basic data input and manipulation can be

delegated and should not require inordinate amounts of time, knowledge or skill.

- Adaptable to disciplines where degree-granting programs or research professionals do not exist at all sites.

- Results can be applied in conjunction with other use data.

- Results can demonstrate a dollar-value in cost reductions if cancellations, rather than substitutions, follow.

- Results yield new insights or confirm anecdotal evidence about collections, their uses, or their users.

THE PROJECT

Assumptions

For the purpose of this study, each author was assumed to retrieve a cited article once, regardless of how many times the article was actually cited. If a title was held at the author's institutional library, the author was assumed to have used the library copy. In a few cases, joint authors from the same university were each credited with a single retrieval of a cited article. This count of unique cites became the surrogate for demand on library collections.

The choice to measure journal citations, rather than citations of monographic serials, was made to focus on the more visible and costly medium for scientific communication. Because publication (or thesis or dissertation deposit) usually occurs several years after the initiation of research, only journals containing cited articles published in 1985 or later were used in this particular study. The database itself, however, can be subsequently mined for usage information on non-current journals. This latter information could be useful for local or Regential preservation, weeding, or off-site storage decisions.

Geologists, like other scientists, cite articles in journals that are targeted to other disciplines. These titles may have a strong local readership, but a measure of their citations by the geologic community would not reflect overall demand on library collections. I assumed that such titles could be readily identified and subsequently excluded from cooperative collection development recommendations until measured against their use by their primary target audience.

Procedure

Using the 1993 edition of the American Geological Institute's (AGI) Directory of Geoscience Departments,

I identified geoscience faculty at ISU, UI, and UNI. Because of an existing agreement, the UI Libraries provide library services for the Geological Survey Bureau (GSB) of the Iowa Department of Natural Resources. A list of GSB professional staff was also obtained. In late April 1994, a letter was sent to each geoscience faculty member (except emeritus and adjunct known to be located off-campus) and each GSB professional. The letters requested photocopies of reference lists from all published works issued from 1991 through 1993. Photocopies of reference lists from grant proposals submitted during that time, regardless of whether or not they were subsequently funded, were also solicited. Academic department chairs were also requested to provide reference lists from student publications issued from 1991-1993 and copies of references from theses and dissertations deposited during those three years. Responses were requested by the end of June. The lists of theses, dissertations, and student publications in the University of Iowa Geology Department's annual Geology Newsletter were consulted, although the periods of coverage were not complete. The ISU Department of Geological and Atmospheric Sciences and the UNI Department of Earth Science provided lists of thesis and dissertation titles.

Because some respondents sent a list of publications, rather than the reference lists from those publications, it was necessary to use interlibrary loan and to enlist the assistance of the UI Geology Library staff to obtain information. As lists were received, I reviewed and marked each for serial titles that were periodical in nature or items in consecutively-numbered series that were rarely or never monographic.

The Library Assistant 3 (LA3) in the ISU Library Collection Development Department organized and input the data into a file using Lotus 1-2-3, version 2.01, which was the only version available to her at the time she began her work. During the Summer, the Department staff were connected to the Ethernet and obtained access to more current versions of Lotus. Because of the size of the file and the press of time, we continued building the file in the older version. The total file contains 2816 unique demands on Regents library collections.

As reference lists were input, the need to standardize titles became clear. Some authors cited the Bulletin of the Geological Society of America, others cited the GSA Bulletin or the Geological Society of America Bulletin. To reduce confusion, titles were standardized using the ISU current form of entry. If ISU had no holdings, the UI form of entry was used. Other authorities used were the *Bibliography and Index of Geology* and the *Chemical Abstracts Service Source Index*. While standardizing titles was beneficial to those working with the database, the process was tedious and required considerable care. In some cases, standardization raised the potential for misunderstanding. All citations to the *Journal of Sedimentary Petrology* were recorded under a combined heading for the new title, even though those uses involved the older title. References to the *European Journal of Mineralogy* and its three predecessors are all recorded under the current title. Because one library held one or more of the predecessors, but not the current title, unfulfilled demand on collections was exaggerated.

To verify that a library had a subscription to current titles, the LA3 searched SCHOLAR, the ISU on-line catalog, and kardex and order records. OASIS, the UI on-line catalog, was available on the ISU Library Staff Server, and it was also checked. Several problems were encountered when subscription cancellations were posted onto OASIS during the project; for the most part, those cancellations are not reflected in this paper. UNISTAR, UNI's catalog was searched by Telnetting. For UNI and UI holdings, their OPAC's served as the final authority on the existence of a subscription.

Subscription prices were compiled for only those titles with cited articles published in 1985 or later. Most of the price information was retrieved from ISU records (still paper-based, at this time), Ebsconet and Faxon. When a subscription price was not found in those sources, the UI Geology Librarian was contacted, if the title was being received there. For several titles assumed not to be gratis, but for which subscription information was not available, an arbitrary price of $25 was assessed. Prices were the latest available, usually for 1994 or 1993.

Data inputting and manipulation required more time than anticipated, and it encroached upon preparations for the oral presentation in October 1994. Thus, there was not time for the LA3 to investigate and learn techniques for producing ranked lists on the computer; those lists were compiled manually.

Preliminary Results

The Regents' Collection

The most heavily cited current titles overall are listed in Table 1; they comprise the titles in greatest demand by the geologic community for research purposes at the three universities. Only four of the thirty-eight titles fail to appear on the *GeoRef List of Priority Journals* (Pierce, L., 1994), the de facto core list for the discipline. Three of those four titles are interdisciplinary, and two are quite regionally focused (*Bulletin of the Centre of Excellence in Geology, Ecology, Journal of the Iowa Academy of Science:*

JIAS, and Paleobiology.)

All but one title are held at one or more Regents'

Table 1. Most frequently cited current titles - the Regents' collection.

TITLE	NUMBER OF CITES
Geological Society of America bulletin	152
Journal of paleontology	133
Abstracts with programs - Geological Society of America	98
American Association of Petroleum Geologists bulletin	97
Journal of sedimentary research. Section A and B	92
Geology	79
Contributions to mineralogy and petrology	76
Journal of the Iowa Academy of Science: JIAS	69
Geochimica et cosmochimica acta	67
Journal of geophysical research; JGR	66
Earth and planetary science letters	59
American mineralogist	56
Science	56
American journal of science	54
Journal of geology	53
Nature	53
Economic geology	44
Soil Science Society of America journal	40
Quaternary research	36
Ground water	36
Canadian journal of earth sciences	30
Sedimentology	29
Journal of the Geological Society, London	26
Paleobiology	26
Journal of petrology	25
Geoderma	24
Journal of hydrology	23
Water resources research	22
Precambrian research	21
EOS	18
Soil science	18
Tectonophysics	18
Mineralogical magazine	17
Ecology	16
Journal of metamorphic geology	16
Tectonics	16
¤Bulletin of the Centre of Excellence in Geology	15
Geological magazine	15

APPROXIMATE COST OF THIS COLLECTION (1 SUBSCRIPTION TO EACH TITLE): **$21,399**

¤ = not held at ISU, UI, or UNI

libraries. Bulletin of the Centre of Excellence in Geology is published in Pakistan. Its use reflects the research interests of two UI graduate students and one UI faculty member.

The UI collection

In Table 2 are the current titles cited by UI authors at least fifteen times. Of these twenty-six titles, only four are not on the *GeoRef List of Priority Journals*. In fact, the same four titles noted on the Regents' collection list appear on this list. Two titles on the UI list are not held at the UI Libraries; *Soil Science Society of America Journal* was canceled in 1980; the *Bulletin of the Centre of Excellence in Geology* was the object of a prolonged, but unsuccessful, attempt to place a subscription. There are eight titles cited heavily by UI, but not ISU or UNI authors. Most of them reflect the UI research emphases on paleontology and Quaternary geology. Although held during the period covered by this study, *Geoderma* was canceled in 1994. Despite some gaps, the UI Libraries collection supports most of the research demand.

Table 2. Current titles most frequently cited by UI authors - the UI collection.

TITLES	NUMBER OF CITES
§Journal of paleontology	132
Geological Society of America Bulletin	91
American Association of Petroleum Geologists Bulletin	88
Journal of Sedimentary research. Section A and B	59
Journal of the Iowa Academy of Science: JIAS	54
Geology	47
Science	44
Abstracts with Programs - Geological Society of America	42
Nature	41
¤§Soil Science Society of America journal	38
Earth and planetary science letters	32
Journal of geophysical research; JGR	30
Quaternary research	30
Journal of geology	29
American journal of science	29
Geochimica et cosmochimica acta	27
§Paleobiology	26
American mineralogist	25
§Sedimentology	25
§Geoderma	24
Contributions to mineralogy and petrology	22
§Precambrian research	21
§Journal of the Geological Society, London	20
Canadian journal of earth sciences	19
¤§Bulletin of the Centre of Excellence in Geology	15
Ecology	15

APPROXIMATE COST OF THIS COLLECTION: $11,797

§ = not appearing on lists of titles most cited by UNI or ISU authors
¤ = not held at UI

Table 3. Current titles most frequently cited by ISU authors - the ISU collection.

TITLES	NUMBER OF CITES
Abstracts with programs - Geological Society of America	56
Contributions to mineralogy and petrology	54
Geological Society of America bulletin	53
Geochimica et cosmochimica acta	44
§Economic geology	41
Journal of geophysical research; JGR	36
Journal of sedimentary research. Section A and B	33
American mineralogist	31
Geology	31
§Ground water	27
Earth and planetary science letters	27
American journal of science	25
§Journal of petrology	21
Journal of geology	19
§Journal of hydrology	17
§Water resources research	14
Nature	12
§Tectonics	12
Journal of the Iowa Academy of Science: JIAS	11
§Journal of metamorphic geology	11
Canadian journal of earth sciences	11
§EOS	10
§Journal of structural geology	10
Science	9
§Tectonophysics	9
American Association of Petroleum Geologists bulletin	9

APPROXIMATE COST OF THIS COLLECTION: $15,378

§ = not appearing on lists of titles most frequently cited by UNI or UI authors

The ISU Collection

The twenty-six current titles most cited by ISU authors covers those cited at least nine or more times (Table 3). Only one title on the ISU list (*Journal of the Iowa Academy of Science: JIAS*) did not appear on the *GeoRef List*. About 40% of the titles are unique to the ISU list. Overall, the list shows a focus on applied and hard rock geology that is complimentary to the emphases at UI. The ISU Library collection is able to fill the research journal demand of its faculty.

The UNI Collection

The University of Northern Iowa is a small institution, and the Department of Earth Science is also smaller than those of the other two Regents' institutions. Table 4 includes all the current titles cited by UNI authors. This list is quite short, and it is very heavily weighted by the research interests of the few respondents. More than half of the eighteen titles cited do not appear on the *GeoRef List*:

> *Annals of the Association of American Geographers*
> *Arctic and alpine research*
> *Canadian journal of botany*
> *Ecological monographs*
> *Ecology*
> *Journal of ecology*
> *Journal of Quaternary science: JQS*
> *Journal of the Iowa Academy of Science: JIAS*
> *Lethaia*
> *Wetlands*

Clearly the demand on UNI Library collections is fulfilled for the most heavily cited titles, save one. *Quaternary Research* should be a top priority for a library subscription, especially if Quaternary studies continue to be emphasized.

Grant Proposals

Only fifteen grant proposal bibliographies were received; current titles cited more than once are shown in Table 5. A subscription to every title in this list is held by at least one Regents' library. Six titles that are not among those mostly frequently cited by Regents' geology authors are unique to this list; most of those titles reflect the growing research emphases on environmental studies, geochemistry, and Quaternary geology. The *GeoRef List* lacks five titles that appear in Table 5:

> *Ecology*
> *Journal of the Iowa Academy of Science: JIAS*
> *Current research in the Pleistocene*

Analytical chemistry
Journal of environmental quality

Except for the one regional title, this group confirms the growing interdisciplinarity of geological research. Despite a small response rate for grant proposal reference lists, the Regents' libraries do appear to be able to support these needs.

Additional Analyses

A list was compiled of current journals cited at least once during 1991-1993 and now costing more than $500 per year. The list was purged of titles whose primary target audience is outside the geologic community. Library subscriptions for remaining titles were compared to the affiliations of those Regents' authors who cited them. Table 6 contains a list of titles, grouped by subscribing institution, that are potential candidates for cooperative collection development. With

Table 4. Current titles cited by UNI authors - the UNI collection.

TITLES	NUMBER OF CITES
Geological Society of America bulletin	8
Journal of geology	5
¤Quaternary research	5
§Ecological monographs	4
Journal of the Iowa Academy of Science: JIAS	4
Science	3
¤§Journal of Quaternary science: JQS	1
§Lethaia	1
§Photogrammetric engineering and remote sensing	1
§Journal of ecology	1
Ecology	1
¤§Arctic and alpine research	1
§Canadian journal of botany	1
Geology	1
§Annals of the Association of American Geographers	1
¤Abstracts with programs - Geological Society of America	1
§Soil science	1
¤§Wetlands	1

APPROXIMATE COST OF THIS
   COLLECTION: **$2,930**

¤ = not held at UNI
§ = not appearing on lists of titles most frequently cited by UI or ISU authors

corroborating circulation and on-site use data, it may be possible to cancel the titles at these particular institutions and local demand satisfied through interlibrary loan from another Regents' library. The UI lacks any expensive titles as potential for cancellation, showing the effects of repeated deep cancellation efforts that were mandated in the last decade. It should be noted that expensive titles held at Regents' libraries, but *not* cited by their authors in the geologic community during 1991-1993, were not revealed by this project. Their discovery must be locally generated. Optimally, such titles would be investigated and considered for

cancellation sooner than would those expensive, but cited titles. The analysis of titles for potential cancellation and cooperation was rendered preliminary by the low faculty response rate described below. In fact, the intended further analysis of titles within the statewide environment for cooperation has been postponed until a better response is received from faculty.

Problems

Although a full coverage of theses and dissertations was achieved, the response rate from faculty was low. Only 40% of faculty and GSB professionals from the UI responded. The rates for faculty response from ISU and UNI were 31% and 38%, respectively. To determine whether these rates would compromise project results, I checked for gaps in faculty research interests, as listed in the 1993 AGI *Directory of Geoscience Departments*. Several major existing research areas were wholly unrepresented by faculty reference lists:

ISU economic geology; most aspects of soft-rock studies; igneous petrology; hydrogeology; and earth science teaching
UNI earth science teaching
UI structural geology; igneous and metamorphic petrology; and geochemistry

Faculty research interests for GSB professionals at the UI were not recorded in the AGI *Directory* except for those geologists who also had faculty appointments. Thus, the gaps in coverage at UI are more extensive than listed above.

One department Chair reported student publications, and data for UI was culled from an annual publication. However, the lack of information from the third school and the overall small size of this population precluded any valid use of these data.

Secondly, there was some difficulty encountered in compiling and manipulating the database. The complexity of the content and the size of the task mandated use of permanent staff. In fact, the LA3 spent approximately 2 months FTE of her time on this project. This necessitated reprioritization of her other work and some rearrangement of the Collection Development Department's clerical work load, generating some complaints in the Department. When the press of regular duties precluded work on the database, delays were encountered. The LA3 experienced difficulties with citations to journals in other languages, especially German and French. Due to the strong UI research interests in paleontology, there were also many citations to nineteenth-century journals which were difficult to

Table 5. Current titles most frequently cited in faculty grant proposals

TITLES	NUMBER OF CITES
Abstracts with programs - Geological Society of America	52
Geological Society of America bulletin	24
Earth and planetary science letters	18
Journal of geophysical research; JGR	17
Quaternary research	16
Nature	16
Contributions to mineralogy and petrology	15
Science	12
§Journal of structural geology	10
Journal of metamorphic geology	10
Ecology	9
Geology	8
American journal of science	8
Journal of the Iowa Academy of Science: JIAS	8
Geochimica et cosmochimica acta	6
Tectonics	5
Mineralogical magazine	4
Water resources research	4
American mineralogist	4
§Chemical geology	3
§Current research in the Pleistocene	3
§Analytical chemistry	3
Journal of petrology	3
EOS	3
§Geophysical research letters	2
Journal of the Geological Society, London	2
Journal of geology	2
§Journal of environmental quality	2
Canadian journal of earth sciences	2

APPROXIMATE COST OF THIS COLLECTION (1 SUBSCRIPTION TO EACH TITLE): **$15,616**

§ = not appearing on list of most frequently cited current titles (Table 1)

Table 6. Expensive titles (>$500) held but not cited; further review recommended.

UI	
None	
ISU	
Earth surface processes and landforms	$635
Palaeogeography, palaeoclimatology, palaeoecology	$1539
Paleontological journal	$596
UNI	
Geochimica et cosmochimica acta	$895
Journal of volcanology and geothermal research	$1155

interpret. Finally, there were several instances when I gave ambiguous instructions, and work had to be redone.

In two circumstances, in-depth subject knowledge of the principal investigator became a factor. In the first instance, the selection of journals from the totality of serial citations would have been a slow and laborious process, requiring examination of actual issues, for someone lacking familiarity with the titles. An alternative approach, to include all serials, would have diminished the value of any results. Another set of judgments was necessitated when the list of titles for potential cooperation (Table 6) was created. In-depth subject knowledge of titles precluded the need to examine issues when deciding which titles were not targeted to the geologic community.

Finally, I am unaware of any recent circulation or on-site use data at any of the Regents' institutions that would validate or repudiate the idea that expensive titles held but not cited are candidates for local cancellation. Noga, Derksen, and Haner suggested that the reliable use data for such a decision would require on-site and circulation data and citation results from theses and dissertations.

NEXT STEPS

Before conclusions are reached about the viability of this methodology, it is necessary to achieve a basic coverage of faculty interests, although responses from every faculty member are not critical. A re-solicitation is planned for those faculty who did not answer the initial letter. In fact, any application of this method will have to allow enough time and staffing for one or more follow-up communications. The other major drawback in this method is its requirement for skilled clerical support. For such a project to reach the analysis stage, it may be necessary for clerical support to be cooperatively funded.

Once results are obtained, they must be communicated to the selectors at the other Regents' institutions. If the method is found viable, then it can be introduced to the Interinstitutional Collection Management Committee, which oversees cooperative projects and the exchange of information among collection development professionals in the Regents' libraries. For the method to be judged truly viable, it should be successfully tested in another discipline with an identifiable core literature and research community.

ACKNOWLEDGMENTS

I wish to thank Carolyn Erwin for database creation and manipulation.

I also thank Leo Clougherty and Nancy Ritchey of the University of Iowa Geology Library for their assistance in obtaining reference lists from UI theses and dissertations and for pricing information.

REFERENCES

American Geological Institute, 1969- , Bibliography and index of geology: Alexandria, Virginia, American Geological Institute.

Chemical Abstracts Service, 1985, Chemical Abstracts Service source index, 1901-1984 cumulative, 2 v.

Claudy, N. H., 1993, Directory of geoscience departments, 32nd ed.: Alexandria, Virginia, American Geological Institute, 466 p.

Noga, M. M., Derksen, C. R. M., and Haner, B. E., 1994, Characteristics of geoscience serial use by faculty and students, *in* Wick, C., ed., Finding and communicating geoscience information: Geoscience Information Society Proceedings, v. 24, p. 61-97.

Pierce, L., GeoRef's List of Priority Journals, American Geological Institute, GEONET-L, 24-May-1994).

THE LITERATURE CITED BY
WATER RESOURCES RESEARCHERS

Richard D. Walker
Myeonghee Lee Ahn

School of Library and Information Studies
University of Wisconsin-Madison
600 North Park
Madison, Wisconsin 53706

Abstract — The availability and use of water resources literature produced by federally sponsored research in general and by the Water Resources Research Institutes in particular was determined. By identifying where the research literature is published, who sponsored it, what publication types are used to record research findings, and to which subject fields within water resources it belongs, the literature used by the water resources community can be characterized. Indicators of quality include whether or not the research has received financial support and the source of that funding. (Financially supported research is a screened subset of the total body of research.) If funding implies quality and funding sources require research findings to be published, it will be useful to know if the target research community uses the published research results.

This study investigated the use of literature as represented by the works cited by U.S. water resources personnel reporting their findings between January 1987 and June 1992 as included in *Selected Water Resources Abstracts*. The differences in use by those supported by U.S. federal funds and those not so funded are determined as are differences in publication types, institutional affiliations, and age of cited literature. Descriptive statistics are also presented to further define the cited literature.

INTRODUCTION

Whether federally funded researchers in water resources cite the same literature to support their research as researchers funded by other sources and whether the literature resulting from federally funded research is used as often as literature of research funded by other sources was investigated. The study focused on utilization of federally funded research in water resources. It is based on the assumptions that knowledge production and utilization are important to water resources research and that the dissemination of knowledge and information resulting from research funded by the federal government is indispensable in meeting the requirements for quality research (Fischer, 1980). It further assumes that the U.S. government technical report plays an important, but undefined or poorly defined, role in the information and knowledge diffusion process.

The research methodology was designed to provide information and a better understanding of:

1. The level of use of federally funded research by the water resources research community

2. The factors contributing to the use of water resources literature, emphasizing that funded by agencies of the U.S. government

3. The characteristics of the water resources literature cited by U.S. authors

Research Sponsored by Federal Agencies

Information generated by public funds, such as that produced through research supported by the U.S. Geological Survey (USGS), provides a need to justify the expenditure in the face of scarce geoscience resources. Information's value is partially defined by the level and kind of use made of it by the community for whom it is produced and by others outside the narrow area in which it was created (Roderer and others, 1983). To build upon what is known and to avoid duplication of research already completed, information produced by research and development (R&D) must be readily and rapidly available and known to the research community at large.

This study addresses the role of federally funded research in the dissemination and transfer of information within the water resources community.

Federally sponsored scientific and technical information (STI) is often perceived as not useful. This perception, especially of the technical report and other gray literature has led to "...the very low level of support for knowledge transfer in comparison to knowledge production [and] suggests that dissemination efforts are not viewed as an important component of the R&D process." In addition, "there are mounting reports from users about difficulties in getting appropriate information in forms useful for problem-solving and decision-making" and "rapid advances in many areas of science and technology...can be fully exploited only if they are quickly translated into further research and application" (Bikson and others., 1984). The kind of translation they address requires multidisciplinary efforts directed at the communication of STI to the appropriate research community with well defined problem solving efforts. The more traditional transfer and communication facilities and mechanisms cannot provide the kind of translation needed. Bikson and others (1984) express the concern that foreign competitors may be better able to apply the results of basic research conducted in the United States and supported by federal research funds than researchers in this country.

The federal government does not have a coherent, centrally located, uniform system to disseminate information produced by the U.S. research community. Individual agencies tailor information dissemination plans to meet their needs with little regard for uniformity or compatibility with other agencies (even those in the same department or parent agency). In addition, the efforts have evolved over time to reflect changing views of administrators, changing agency missions, available budgets, and updated priorities.

The USGS Water Resources Division, manages the state WRRI programs as did its predecessors, the Office of Water Resources Research (OWRR), the Office of Water Resources and Technology (OWRT), and the Office of Water Policy (OWP); an information transfer system including the Water Resources Scientific Information Center (WRSIC), is designed to acquire, process, announce, abstract, index, and publish the results of federally funded research, and research in any publication format from all sources, including open-file and government technical reports.

Communication of Scientific and Technical Literature

Any literature review of the communication or transfer of STI reveals that research communities rely on two major systems for knowledge and the acquisition of STI. The two systems consist of the formal and the collegial networks (Cronin, 1982; Gerstenfeld and Berger, 1980). Even when they are both functioning well, gaps remain. The formal system tends to identify larger numbers of sources, published documents for the most part, while collegial contacts complement the formal system with information and informed judgements about less well known sources. The formal system tends to work best with older materials and the collegial contacts provide information on more recent work. The only true measure of the value of the systems is one that assesses the results as measured by the communication of information to those who need it.

Many federal agencies support research in areas where researchers prefer the scientific journal as the medium for addressing their audiences; technical reports are produced only as a contract requirement. Mission-oriented agencies such as the U.S. Department of Energy (DOE), the National Aeronautics and Space Administration (NASA), and the USGS contract with universities, large private organizations, and small businesses to perform research. The focus of the research is on specific problems, with the technical report serving as the mechanism for recording and disseminating results and meeting contractual obligations. Because it contains greater detail the technical report is often the best repository for information on procedures and processes in addition to the recording of results.

The federal government plays a dominant role in the production of STI. Rosenberg (1985a) stated: "The role of the Federal government in the support of R&D is carried out within an institutional framework dominated [or characterized] by contractual relationships between the Federal government, the aerospace industry, and the academic community." Similarly, as Pinelli (1991) observed, "These contractual relationships, in and of themselves, contribute to the transmission and utilization of knowledge resulting from federally funded aerospace R&D." He further observed that the transfer of knowledge is aided by joint government-industry cooperative projects, by groups composed of representatives from universities, government and industry, and by the use of government facilities by universities and industry, the exchange of personnel for research, and jointly sponsored workshops and conferences.

Many federal agencies have STI systems for acquiring, processing, announcing, publishing and otherwise transferring government-performed and government-sponsored R&D. They are the primary means of transferring STI resulting from government contracts with academia or industry to the broader research community.

The role of the federal government in STI evolved with the recognition that change was necessary if government agencies were to accomplish their missions in a timely manner. Therefore, selective activities were supported as a part of the federal funding for R&D. Support for STI was often associated with achieving foreign and military policy objectives, but gradually *ad hoc* support evolved into permanent relationships with the research community, e.g., the establishment of the National Academy of Sciences (NAS) in 1863 and the USGS in 1879.

Ballard and others (1989) point out that as the federal role in R&D increased, so did its presence in technology transfer. They cite the creation of the U.S. Cooperative Extension Service in 1914 as an example of an earlier, highly successful technology transfer program supported by the federal government. This service included the land grant universities in the dissemination of new agricultural practices, the meeting of information needs of a wide range of users, and the increased availability of research findings. Feller (1986) notes that this national partnership between the federal government and researchers is credited for much of the technological progress of the American agricultural system. Elsewhere this partnership is credited with contributing to the success of agriculture in this country (Feller, 1987).

The Knowledge Utilization Model of STI transfer is currently the dominant model. Grounded in the diffusion of innovation (Rogers, 1963), the planned change literature (Bennis and others, 1976; Glaser and others, 1983), and the model of social research (Archibald, 1970; Burgess, 1973; Patton, 1978), the Knowledge Utilization Model emphasizes the process by which new ideas are transferred to potential users.

RATIONALE FOR THE STUDY

It is essential to understand how scientific and technical information is communicated in the overall process of technological innovation and scientific advancement and to comprehend how and where this communication takes place. Broad federal polices that influence the production, transfer, and utilization of STI need to be understood. As the role of STI is central to the innovation process, the more we appreciate the process the better we can contribute to it and facilitate it.

Much can be learned about a discipline from an analysis of its literature. A discipline's literature identifies the problems under study and becomes the knowledge base for the field. From literature analysis one can learn what is being researched, who is doing the research, and who is using the research results to continue in the same direction or pursue another track. By studying who and how research is sponsored one can gain insights into what is valued or deemed important by the research community and society at large. Over time the evolution of a discipline can be ascertained and known through the literature analysis. Shifts in direction or even its knowledge base can be determined, as can the intended audiences.

The central role of STI in the efficient conduct of research is well established. Others have found relationships between the communication of STI and technical performance (Carter and Williams, 1957; Allen, 1970; Rosenbloom and Wolek, 1970; Rubenstein and others, 1971; Hall and Ritchie, 1975). Fischer (1980) concluded that the "role of scientific and technical communication is (thus) central to the success of the innovation process, in general, and management of R&D activities, in particular."

Using a variety of approaches and motivations, bibliometricians have built up a body of literature to assist in understanding many disciplines, particularly in science and engineering. Pao (1991) found that funded authors were cited more often than unfunded authors. Others too have researched the relationship of funding to use and quality.

Bikson and others (1984) identify five major problems in assuring the availability of STI:

1. Many potential users are unaware of much of the information available to them
2. Many potential users are discouraged by the complexity and cost of obtaining information from abstracting and indexing services
3. Dissemination of research products is narrow
4. Formal information transfer systems are not timely enough
5. The continuity of abstracting and indexing systems is threatened by budgetary constraints.

A journal's reputation and the reputation of the publishing society are generally thought to provide a high level of control for their published articles (Abelson, 1980), although some critics are skeptical of peer review. Technical reports do not undergo the same kind of peer review as journal articles but work is often critically judged before contracts are let and during the period of research (Passman, 1969; Walker and Hurt, 1990).

A possible indicator of quality of research may be the funding it received. Research projects are selected for funding from a pool of research proposals. It might be assumed that the act of selecting a project for funding by an agency implies a certain worth to the proposal. The reasons for selection might be varied, from an anticipated contribution to the field of study, a solution to a known problem, a demonstration of a model or process, the new application of a research method, or the development of a research tool, such as a computer program or formula.

Related Literature

Citation counts are employed as a measure of usefulness of research output. The appropriateness of citation analysis for evaluation is demonstrated by research that correlates citation counts with other measures of quality (Anderson and others, 1978; Koenig, 1983; Lawani and Bayer, 1983; Narin and others, 1976).

Harter and Hooten (1990, 1992) studied the relationship between funding and quality, testing whether information science articles reporting the results of funded research were better or more useful than articles reporting unfunded but similar research. In 1990, they examined variables associated with articles published in the *Journal of the American Society for Information Science (JASIS)* from 1972, 1973, and 1974. They found no significant relationship between funding status and the number of citations it attracted, i.e., no evidence that funding has anything to do with how often *JASIS* papers are cited. They did find a significant relationship between the institutional affiliation of the first author and the funding of the reported research and also between institutional affiliation and the number of citations attracted. In 1992, nine volumes of the same journal for the years 1972-1974, 1982-1984, and 1988-1990 were used to test differences among six variables from each full-length article. No relationship was found between whether an article was funded and the quality or utility of the article, as measured by the number of subsequent citations to it.

Studies have shown the importance of interpersonal channels to STI communication, two examples are Isenson (1968) and Bernal (1967). Networks of researchers have been shown to be critical to the advancement of research (Crane, 1972) because they have the advantage of being immediate and subject to quality control. They also permit feedback as needed (Fischer, 1980). Scott (1962) found that literature does not serve so much as a reference source as it does a source of stimulation for further research:

> The principal role of literature is to supply useful information which is not being deliberately sought.... Any approach which takes for granted that the reader of technical literature is typically engaged in a search for a particular piece of information is seriously out of touch with reality.

METHODOLOGY AND RESEARCH DESIGN

A citation analysis of a sample of the research indexed in Selected Water Resources Abstracts (SWRA) was used to conduct a bibliometric analysis of the artifacts of water resources research as recorded in the cited documents by U.S. researchers. Author affiliation, subjects researched, federal funding status, agencies funding the research, and the publication types used to report research were recorded.

A random sample of 400 documents, the source documents for the citation study, was drawn from >39,000 documents published from January 1987 to July 1992 and indexed in SWRA, available as a CD-ROM by the National Information Services Corporation, Baltimore, Maryland (NISC/DISC). This was a randomly selected sample of the most recent research literature available at the beginning of the project and all publication types used by U.S. authors to report research findings were included. Only documents with at least three cited works were defined as research and included in the sample. Review articles were excluded (Ahn, 1993).

Documents making up the sample were the source or citing documents for this investigation. For each citing document the following information was recorded:

1. Funding status of the research (and if with federal funds, which agency)
2. Publication type, i.e., journal article, technical report, conference proceeding, thesis, book or part of book, government monograph, or others
3. Institutional affiliation of first author, i.e., university, commercial, government, or non-profit
4. Subject within water resources, as identified in SWRA.

For each work cited by the source or citing document (i.e., endnote or footnote) the following information was recorded:

1. Age in years (publication date of cited document subtracted from publication date citing document)
2. Publication type (same as above)
3. Institutional affiliation of first author
4. Country of residence of first author
5. Language of the document
6. Funding status (federal or other)
7. If federally supported, by which agency.

There are nearly 9,500 cited documents generated

Table 1. Cited literature funded by federal and nonfederal sources.

| Research Support for Citing Source | | | | | |
| Non-Federal | | Federal | | Total | |
Frequency	Percent	Frequency	Percent	Frequency	Percent
Non-Federally Funded Cited Documents					
1,480	21	1,166	16	2,646	37
Federally Funded Cited Documents					
1,872	26	2,574	37	4,446	63
Total					
3,352	47	3,740	53	7,092	100

from the 387 source documents, or an average of 25 citations for each citing document. Thirteen source documents initially selected in the sample were not used in the investigation because they could not be obtained. A small number of cited documents were not examined because they were unobtainable or the citations were so incomplete that they could not be identified. Efforts were made to locate copies of publications not available in the Library System at the University of Wisconsin-Madison using interlibrary loan, direct contact with distributing sources, and appeals to authors or source institutions. Attempts to identify improperly cited documents were made with some success.

The statistical tests used to determine differences are Chi-square, t-test, Savage Scores (exponential) test and Scheffé comparisons. All hypotheses are tested for significance at $\alpha = 0.05$.

STATISTICAL ANALYSIS

The primary hypothesis tested was that U.S. funded and other research is cited differently by authors with federal support than those without such support. The null hypothesis is that no difference exists in the level of cited research between the two groups of researchers. In addition, for each citation, the following information was collected:

1. institutional affiliation
2. publication type
3. whether U.S. federal funding results in differences in the body of cited documents.

To test for differences a Chi-square contingency table was used to see the relationships between funding status of cited documents and citing documents (Table 1). Overall, U.S. authors cite more federally funded research (63%) than non-federally funded research (37%), and slightly more than half of the authors (53%) are supported by federal funds. More precisely, federally supported authors cite federally funded research twice as often as non-funded research, while non-federally supported authors cite slightly more funded research (26%) than non-funded research (21%).

From the Chi-square test, with $p = 0.0001$, the null hypothesis that no difference exists between the two groups of researchers is rejected and the research hypothesis was accepted. Federally funded researchers cited more federally funded research than authors funded by other sources. This hypothesis was tested using the entire population of source documents without regard to institutional affiliation, publication type, or subject within water resources. All authors of the source documents were from the United States. Publication type and institutional affiliation were tested separately.

Type of Publication

The second hypothesis to be tested was whether federally funded and other research is cited differently among authors publishing in journals, technical reports, conference proceedings, books or parts of books, government monographs, theses and other types of publications (e.g., maps, patents, and guidebooks) (Table 2). The null hypothesis tested was federally funded research is cited no differently than research funded by other sources by authors publishing in various publication types. Type of publication was recorded to determine if water resources researchers favored certain types of publications more than others. Of the 7,092 cited documents 53% were funded while 47% were not. Fifty-four percent of citing documents are journal articles, books (17%), technical reports (12%), conference papers (6%), monographs (5%), theses (3%) and others (3%). This indicates that federally funded

Table 2. Cited water resources research designated by funding source and publication type.

Cited Research	Journal	Technical Report	Conference Proceeding	Book	Theses	Monograph	Other	Total
Funded								
Frequency	2,052	670	210	372	66	243	127	3,740
Percent	29%	9.4%	2.9%	5.2%	0.9%	3.4%	1.8%	53%
Non-funded								
Frequency	1,767	171	196	859	130	118	111	3,352
Percent	25%	2.4%	2.8%	12%	1.8%	1.7%	1.6%	47%
Total								
Frequency	3,819	841	406	1,231	196	361	238	7,092
Percent	54%	12%	5.7%	17%	2.7%	5.1%	3.4%	100%

research was cited differently than research funded by other sources by authors publishing in journals, technical reports, conference proceedings, books, government monographs or theses. Among the six types of citing documents, the number of technical reports affected the results the most: Authors publishing in technical reports cite more funded research than other research.

Institutional Affiliation

The third hypothesis tested was whether authors with different institutional affiliations cite different types of literature. From Table 3 it can be seen that slightly more than half of the documents cited (53%) were federally funded and 47% were funded by other sources. Of the citing authors 52% were affiliated with universities, 31% with government, 14% with commercial firms and 2% with the non-profit sector.

Results indicate that federally funded research is cited differently than research funded by other sources by authors in university, government, commercial firms, and the non-profit sectors. In particular, authors affiliated with the government exhibited the most varied citation patterns.

Age

The mean age of literature cited was tested to determine if federal support affected the age of the literature. Age was determined by subtracting the year of publication of the cited document from the year of publication of the citing document, (e.g., if the citing article was published in 1990 and cited a document published in 1965 the age was recorded as 25 years).

To test mean differences of literature cited between authors of federally supported research and authors funded by other sources, Savage Scores were used. The mean age of literature cited by authors having federal support was 11 years, while that cited by authors

Table 3. Water resources researchers: Their funding sources and institutional affiliations.

Cited Research	University	Government	Commercial	Non-Profit	Total
Funded					
Frequency	1,803	1,378	454	105	3,740
Percent	25%	19%	6.4%	1.5%	53%
Non-funded					
Frequency	1,907	871	513	61	3,352
Percent	27%	12%	7.2%	0.9%	47%
Total					
Frequency	3,710	2,249	967	166	7,092
Percent	52%	31%	14%	2.4%	100%

Table 4. Cited research in water resources: Its age and publication type.

	Journal	Technical Report	Conference Proceedings	Book	Theses	Monograph	Others	Total
Number	5,463	904	490	1,564	218	410	258	9,043
Mean Age	11	9.1	7.8	12	8.5	21	24	11
Standard Deviation	11	10	7.8	12	8.4	24	43	12

supported by other sources was 10 years. These data are close to an exponential distribution, which does not permit the reliable use of the t-test. With an exponential distribution the Savage Scores (exponential) test is more powerful. Using the Savage Scores (exponential) test the null hypothesis is rejected at $p=0.0001$. Thus, the hypothesis that the mean age of literature cited by authors is different between those who received federal support and those not so funded is accepted. More precisely, the mean age of literature cited by authors who have federal support is older than that cited by authors not so funded.

Age And Publication Type

Statistical analysis of the data was used to determine if the age of the literature cited by authors varied according to the type of publication (i.e., journal, technical report, conference proceedings, book or chapter, government monograph, thesis) in which research results were published. Table 4 shows the mean age of the literature appearing in various publication types. Results indicate that the mean age of cited literature differed among the publication types.

Even though the Savage Scores Statistics show differences among the means, it does not tell which mean is significantly different among the six. To investigate which means are different among the six publication types the Scheffé comparison is used as a *post hoc* test. The Scheffé comparison showed there are significant differences between conference proceedings and journals, conference proceedings and technical reports, conference proceedings and books, and conference proceedings and theses and books and theses.

Age and Institutional Affiliation

To determine if the age of literature was affected by the institutional affiliation of the researcher the mean age of literature cited by authors from university, government, commercial and non-profit sectors was compared. As Table 5 shows, the average mean age of literature was 11 years. Mean age of literature cited by authors employed in government was 13 years, universities 11 years, commercial firms 10 years and non-profit organizations 10 years. It was found that the mean age of literature cited by authors differed with institutional affiliation.

As with publication type, the Savage Scores Statistic does not show which means differ among the four means. To identify which means were significantly different, Scheffé comparisons were used. The Scheffé comparisons showed significant differences between authors in government and commercial firms and between authors in government and non-profit groups.

DESCRIPTIVE STATISTICS

As the data to be statistically tested were collected additional characteristics recorded were:

1. country of origin of the literature based on the institutional affilitation of the first author
2. language of the cited document
3. agency from which the federal funds were received, when applicable
4. subject assigned to each of the citing documents by SWRA.

These characteristics are discussed in an effort to further describe the water resources literature cited by U.S. authors.

Table 5. Age of cited water resources research and the institutional affiliation of the researchers.

	University	Government	Commercial	Non-Profit	Total
Number	5,040	2,870	1,179	218	9,307
Mean Age	11	13	10	10	11
Standard Deviation	12	18	11	10	14

Country of Origin of Reporting Researchers

Based on the data collected, the majority of the cited documents originated in the United States. This was expected because, as with other geosciences, water resources literature reflects the geographical area where the research is performed. Of the 9,292 documents for which there were data, 7,100 (76%) were reported in English by U.S. authors; 557 (6%) by Canadian authors; and 444 (5%) by British authors. Others included: 156 from Germany; 79 from France; 149 from Australia; 97 from the Netherlands; 95 from Japan; 48 from New Zealand; 173 from Scandanavia; 50 from Russia; 49 from countries that were a part of the USSR; 118 from all other European countries; 60 from the Middle East; 51 from Asia (exclusive of Japan); 22 from Latin America; and 44 from Africa.

Language of Cited Literature

Nearly all of the literature was reported in English language. Clearly, U.S. researchers in water resources use the English language literature almost to the exclusion of other languages. One reason for this is that the research is mainly concentrated in the United States. Of the 9,304 documents with data on language, 9,248 (99%) were in English, 22 were in French, and 13 were in German. Dutch and Spanish were each used twice, Russian once, and Scandinanian languages five times. No other language was used more than five times. A total of 10 languages was used in the 9,304 cited documents. Only a very few pieces of foreign language literature was cited by permanent residents of the United States and then only when the area treated was outside the United States.

U.S. Agencies Supporting Water Resources Research

The distribution of agencies funding water resources research is broad, but concentrated within a few agencies. The USGS and USEPA were most frequently reported as funding agencies. Of the 9,315 documents examined in this study 3,803 resulted from funded research and 5,512 were the result of unsupported research. The USGS funded 741 of the projects while the USEPA supported 747 of the projects; a total of 40% of the funded projects. Other funding agencies included: National Science Foundation (NSF), 525 documents (14%); USDA, 362 documents (10%); USDOE, 262 documents (7%); other U.S. Department of the Interior (USDI) programs, 248 (7%); National Oceanic and Atmospheric Administration (NOAA) with the U.S. Department of Commerce (USDC), 165; the U.S. Army Corps of Engineers (COE), 125; NASA, 25; and the National Academies of Science and Engineering, 33. There were 370 documents resulting from research that was supported by more than one agency.

Subjects of the Citing Documents

Of the 9,307 documents cited for which subjects were recorded, 3,561 (38%) were identified as contributing to the study of the water cycle; 4,664 documents (50%) were assigned to water quality management and protection; engineering works, 331 documents (4%); water quantity, 292 documents (3%); water resources and water resources planning with 172 (2%) and 174 documents (2%), respectively; water supply, 112 cited documents, or just over 1% of the total.

Although the authors did not examine the cited literature that falls outside water resources, the following observations were made: Little literature outside of water resources is cited, and much of the literature that falls outside water resources consists of methodological techniques and procedures, computer programs of broad application, and statistical programs and techniques of broad application.

SUMMARY AND CONCLUSIONS

The cited documents are produced in this country, written in English, mostly from the field of water resources and slightly over half (54%) are journal articles. There is also a large body of literature reported in technical reports, government monographs and conference proceedings. A larger than expected amount of literature is found in books, cited either as an entire book or part of a book. Although there were not many theses in the sample a relatively large number of references cite theses (196 or 2.8%). This indicates that theses and dissertations are viewed by the water resources research community as an important information source and should continue to be included in the SWRA.

It was concluded that researchers funded through one or more federal agencies were more likely to cite federally funded research than were other researchers. At the same time, funded research represents 63% of the literature cited by all researchers. Researchers employed in a federal agency cite more federally funded research than those employed elsewhere. Researchers published in government monographic series cite more federally funded research than those publishing in other publication types and academic researchers are the heaviest users of federally supported research and nonsupported research.

It is fair to say that the source of support affects the

use of the literature. Researchers who received support from the federal government make heavier use of federally supported research than researchers not receiving support. As noted above, United States authors cited almost twice as many federally funded research documents as those not funded. As the same time over half of citing authors are supported by federal funds and even non-federally supported authors cite slightly more funded- than non-funded research. The findings show that federally funded research – as a major source of water resources literature – is highly regarded by the research community.

The literature is cited differently by researchers who publish their findings in different publication types, with authors of technical reports citing more funded research than nonfunded research.

There is also different use of the literature by authors with different institutional affiliation. We do not know what to conclude from these findings except to note them and guess that the availability of materials in libraries and greater access to bibliographic apparatus in universities as contrasted to the commercial sector may be the cause. The *modus operandi* of researchers and the motivation for research are also reflected in the literature cited to support ideas and discoveries and these vary with place of employment. Government researchers have different work habits and research objectives from those found among university faculty researchers and non-profit and commercial sector researchers.

The age of the literature used by researchers is different for researchers supported by federal funds than those not funded. Authors with federal support use older literature than do authors not so funded. Again, this is difficult to explain. It may be that the availability of materials contribute to this difference and that there are wide differences in the availability of materials among institutional types.

Overall there are differences between the federally funded literature and the literature produced without such funding, although differences are not great. The important finding is that the literature generated by research supported by federal funds is just as highly regarded as that not so supported, but that the format or publication type contributes to its use. One way to increase the use of the research funded by the federal government is to encourage and facilitate publication of research in journals after it is first published in a technical report or thesis. Government monographs are more widely distributed than technical reports and their use should be encouraged. The open-file report, a type of technical report and a format used by many agencies as an alternative publication type, should be discouraged and used only when other types of publication are not feasible. It is the single most fugitive publication type used to record research findings. Many federal agencies make heavy use of this format and the USGS is one of the heaviest users. Not only are open-file reports the most fugitive of publication types, but one of the most difficult to use after identified and obtained by a user. They are now distributed by the USGS through their Denver office in microfiche only, while retaining the only hard copy in their files in USGS offices across the country. In addition, not all open-file reports are distributed in any format, further contributing to the difficulty of access.

Efforts to make federally funded research available should be continued both through distribution and announcement services but with wider availability of the already existing bibliographic control services. Any changes in abstracting and indexing services designed to control the research literature of water resources should be directed toward broader coverage of journal titles and publication types and wider distribution among the researchers and other users.

ACKNOWLEDGEMENTS

This research was supported by the United States Geological Survey, Department of Interior, under a USGS award through the University of Wisconsin, Water Resources Center

REFERENCES

Abelson, P.H., 1980, Scientific communication: Science, v. 209, p. 60-62.

Ahn, M.L., 1993, Retrieval effectiveness of subject descriptor and citation searching in the water resources literature [Ph.D. Dissertation], Library and Information Studies, Madison: University of Wisconsin-Madison, 225 p.

Allen, T. J., 1970, Roles in technical communication networks; in:. Nelson, C.E. and Pollack D.K. eds., Communication among scientists and engineers: Lexington, Massachusetts, D.C. Heath,. p. 191-208.

Anderson, R., Narin, F., and McAllister, P., 1978, Publication ratings versus peer ratings of universities: Journal of the American Society for Information Science, v. 29, p. 91-103.

Archibald, K.A., 1970,Three views of the experts's role in decision making: Systems analysis, incrementalism, and the clinical approach. Policy Sciences 1, p. 73-86.

Ballard, S., Devine, M.D., James, T.E. Jr., Malysia, L.L., Adams, T. I., and Meo, M., 1989. Innovation through technical and scientific information: Government and industry cooperation. New York,

Quorum Books, 38 p.

Bennis, W.G., Benne, K., Chin, R., and Corey, K., 1976, The planning of change, 3rd Ed.: New York, Holt, Rinehart and Winston, 517 p.

Bernal, J. D., 1967, The social function of science. Cambridge, Massachusetts, MIT Press, 482 p.

Bikson, T.K., Quint, B.E., and Johnson, L.L., 1984, Scientific and technical information transfer: Issues and options: Santa Monica, California, Rand, N-2131-NSF, NSF/PRA-84015 and PB 85-150357).

Burgess, P.M., 1973, On putting our oars in the water: A clinical perspective on the design and analysis of Policy change. Lake Cumberland, Kentucky, Foreign Policy Research Conference,

Carter, C.F., and Williams, B.R., 1957, Industry and technical progress: Factors governing the speed of application of science: London, Oxford University Press, 244 p.

Crane, D., 1972, Invisible college: A diffusion of knowledge in scientific communities: Chicago, Illinois, University of Chicago Press, 213 p.

Cronin, B., 1982, Invisible colleges and information transfer: A review and commentary with particular reference to the social sciences. Journal of Documentation, v. 38, p. 212-236.

Feller, I., 1986, Universities and state governments: A study in policy analysis. New York, Praeger, 170 p.

Feller, I., 1987, Technology transfer, public policy, and the Cooperative Extension Service - OMB imbroglio: Journal of Policy Analytical Management, v. 6, p. 307-327.

Fischer, W.A., 1980. Scientific and technical information and the performance of R&D groups, in Burton, V.D., and Goldhar, J.L., eds. Management of research and innovation: New York, North Holland, p. 67-89.

Gerstenfeld, A., and Berger, P., 1980, An analysis of utilization differences for scientific and technical information management science: Management Science v. 26, p. 165-179.

Glaser, E.M., Abelson, H.H., and Garrison, K.N., 1983, Putting knowledge to use: San Francisco, California, Jossey-Bass, 636 p.

Hall, K.R., and Ritchie, E., 1975, A study of communication behavior in an R&D laboratory: R&D Management, v. 5, p. 243-245.

Harter, S.P., and Hooten, P.A., 1990, Factors affecting funding and citation rates in information science publications: Library and Information Science Research, v. 10, p. 263-280.

Harter, S.P., and Hooten, P.A., 1992, Information science and scientists: JASIS, 1972-1990. Journal of the American Society for Information Science, v. 43, p. 583-593.

Isenson, R. S., 1968, Technological forecasting lessons from project hindsight, in: Bright, J.R., ed., Technological forecasting for industry and government: Englewood Cliffs, New Jersey, Prentice-Hall, p. 35-57.

Koenig, M.E.D., 1983, Bibliometric indicators versus expert opinion in assessing research performance: Journal of the American Society for Information Science, v. 34, p. 136-145.

Lawani, S.M., and Bayer, A.E., 1983, Validity of citation criteria for assessing the influence of scientific publications: New evidence with peer assessment. Journal of the American Society for Information Science, v. 34, p. 59-66.

Narin, F., Piniski, G., and Gee, H., 1976, Structure of the biomedical literature: Journal of the American Society for Information Science, v. 27, p. 24-45.

Pao, M., 1991, On the relationship of funding and research publications: Scientometrics, v. 19, p. 199-206.

Passman, S., 1969, Scientific and technological communication: Oxford, Pergamon Press, 151 p.

Patton, M.Q., 1978, Utilization-focused evaluation: Beverly Hills, California, Sage Publications, 304 p.

Pinelli, T.E., 1991, The relationship between the use of U.S. Government technical reports by U.S. aerospace engineers and scientists and selected institutional and sociometric variables: Report No. 6, NASA Technical Memorandum 102774. NASA/DOD Aerospace Knowledge Diffusion Research Project, Washington, D.C, NASA and Department of Defense.

Roderer, N.K., King, D.W., and Brouard., S.E., 1983, The use and value of Defense Technical Information Center products and services: Alexandria, Virginia, DTIC.

Rogers, E.M., 1962, Diffusion of innovations, 2d ed.: New York, Free Press of Glencoe, 367 p.

Rosenberg, N., 1985a, A historical overview of the evolution of federal investment in research and development since World War II: Paper commissioned for a workshop on The Federal Role in Research and Development, November 21-22, 1985, held in Washington, D.C. and sponsored by the National Academy of Sciences, National Academy of Engineering, and the Institute of Medicine, Washington, D.C.

Rosenberg, N., 1985b, The impact of technological innovation: A historical view; in Landau, R., and Rosenberg, N., eds. The positive sum strategy: Harnessing technology for growth: Washington, D.C., National Academy Press, p. 17-32.

Rosenbloom, R.S., and Wolek, F.W., 1970, Technology and information transfer: A survey of pace in industrial organizations: Boston, Massachusetts, Harvard University, 174 p.

Rubenstein, A.H., Barth, R.T., and Douds, C.F., 1971, Ways to improve communications between R&D groups: Research Management, v. 14, (November) p. 49-59.

Scott, C., 1962, The use of technical literature by industrial technologists: IRE Transactions on Engineering Management, v. 9, p. 76-81.

Walker, R.D., and Hurt, C.D., 1990, Technical reports, *in* Walker, R.D., and Hurt, C.D., eds., Scientific and technical literature: An introduction to communication forms: Chicago, Illinois, American Library Association, p. 103-128.

IMPROVING BIBLIOGRAPHICAL ACCESS TO PUBLISHED GEOLOGIC MAPPING BY USING ONLINE MAP INDEXES

Nancy L. Blair

U.S. Geological Survey Library
345 Middlefield Road, M.S. 955
Menlo Park, California 94025

Abstract — Locating published mapping in earth science literature would be simplified by the availability of bibliographical databases which would allow the searcher to use maps to define areas for which mapping on any subject is needed or to search by standard mapping quadrangles. During the past months, selected computer software programs have been compared using a sample geologic map index database to find the best means of combining detailed bibliographical records for maps with geographic access and sorting by scale, date, and other fields of the citation. Other selection criteria included cost, availability and compatibility with other computer environments so that retrieval data could be easily transported into another program. Once software has been selected , the prototype geologic map index program will be used to set up related databases for locating publications on many other subjects such as soils mapping, geophysical mapping, radiometric dating, trace element sampling, and well logs.

INTRODUCTION

Last year in a GSA poster session, Connie Manson of the Washington State Division of Geology and Earth Resources and I recommended a cooperative effort to build a national database for published geologic mapping (Blair and Manson, 1994). Since this is not a reality, the reference librarians of the U.S. Geological Survey in Menlo Park, California, could use immediately a better index to California geologic mapping to adequately answer questions from the public. Without a comprehensive index, several printed and digital bibliographies must be searched which either takes up the librarian's time or requires frequent intervention by the librarian when users do their own searches.

To begin developing a good geologic map index for California, several decisions had to made to insure that the work involved did not exceed its value, that any compilation would migrate easily into other digital environments and not be lost when other options became available, and that a truly workable, easy index with graphical display would be the end result.

Three parts of setting up the index database were investigated: determining the easiest method of locating and entering bibliographic information, selecting the software for recording citations, and choosing the software to produce a visual representation of the boundaries of indexed maps.

LOCATING AND ENTERING BIBLIOGRAPHIC INFORMATION

At first, the fastest way to compile a California geologic map index seemed to be the use of a single bibliographic source such as GEOREF or GEOINDEX augmented with citations from one or two other sources. However comparison searches showed that no single or combination of two bibliographic sources yielded nearly complete coverage of mapping for areas in California. While there would be considerable overlap with duplicate citations, a percentage of retrievals was unique to each bibliographic source. There was no automatic way to remove the duplicate citations or to merge searches from different databases or published sources.

Keyword or subject indexes often missed important maps which used broad geographical terms in the title. These should not have been missed by geographical coordinate searching in a comprehensive digital databases such as GEOREF, but many citations were not indexed using latitude and longitude fields.

Frequent discrepancies between the same citation as entered in different bibliographic sources were found which could only be resolved by looking at the publications to determine if the discrepancies were entry errors or differences in choices such as between the title on a map and that on an envelope. Older citations in the same bibliographical sources were often inferior to more recent entries lacking crucial fields such as coordinates. In the U.S. Geological Survey's (USGS) library catalog, more than three authors were cited as "first author ....[et al]" according to library cataloging rules while USGS citation style asks for listing of all authors requiring recourse to the publication to fill in this information. In GEOINDEX, early classic publications were over-indexed with different map plates in a single publication given as separate entries, causing multiple retrievals of the same publication.

The frequency of these problems showed that, while good databases such as GEOREF or GEOINDEX were assets, the citations could not be used without considerable editing. The fastest and most accurate method was to start from the publications entering the bibliographical infor-mation directly and then using the established bibliographic sources for cross-checking entries. Since the majority of publications were USGS series and California Division of Mines and Geology publications, the entries were entered directly series by series. Obtaining correct information for theses and the limited number of outside publications requires more ingenuity and searching. Determining the coordinates of maps in theses not owned by the library remains a problem.

BIBLIOGRAPHIC SOFTWARE

In the early stages of the computer technology, selecting software and the platform to be used for any extensive work was a crucial question. Without software compatibility, a great deal of work would be unavailable to others or was lost when better software became available. However this is no longer a problem. IBM and Macintosh data files can be translated back and forth. Software allows easy migration from one version of a software to an upgrade and between related types of software designed by different companies such as spreadsheets, relational databases and even GIS databases (Montgomery, 1994). Network software allows communication between different types of computers. A Paradox relational database can be exported as ASCII, dBase, or other files and potential users preferring other software are not isolated from the database information.

Several types of software could work for entering bibliographic entries: word processing programs, spreadsheets, relational databases, and specialized bibliographic software such as *End-note*, *ProCite*, and *Papyrus*. Of these, relational databases such as *dBase*, *Paradox*, and others are the most versatile for this purpose. A relational database permits adding, deleting, and altering fields to adjust the database information for special applications and allows for automatic sorting on all fields. For example, a work field can be added to

record the progress of checking a citation or its origin and removed with a few keystrokes when the database is complete. The order of fields in output can be altered to adjust for differing bibliographical styles. Various sortings by series, author, and title facilitate checking for consistency of entries, possible duplications and typographical errors while constructing the database. Common relational database programs permit exporting to other such databases, to selected specialized bibliographic software programs, and to word processing programs without affecting the compilers' database.

Experimenting has demonstrated the following structure to be adequate for all citations recorded on California geologic mapping with A representing alphanumeric fields, N for numerical fields, and num-bers indicating the spaces needed for information in that field.

1.	Publication code	A	25
2.	Author code	A	40
3.	Author(s) & date	A	110
4.	Year for date sorting	N	–
5.	Title	A	220
6.	Publisher	A	60
7.	Series or imprint	A	60
8.	Collation	A	40
9:	Scale	N	–
10:	Counties	A	60
11.	Quadrangles	A	60
12.	South latitude	A	6
13.	North Latitude	A	6
14.	East Longitude	A	7
15.	West Longitude	A	7
16.	Remarks	A	60

Additional fields are entered to record progress in checking citations and to mark maps which are irregular and require special handling in a graphics program. These additional fields would be removed from any outputs of the database.

The total spaces required for each citation is in excess of 770 spaces. The total number of citations should reach a minimum of 3000 entries.

GRAPHICAL DISPLAY

Ideally the graphical display should be directly linked to the bibliographical database. The software should be able to automatically translate latitude and longitude fields into right-angle polygons for standard quadrangle mapping and permit easy entry of the outlines of maps covering irregular areas.

Eventually to be fully versatile and compatible

with other mapping databases in the organization, the map outlines for a geologic map index should be entered in a GIS (Geographical Information System) system (Larsgaard, 1992) with ARC/INFO being the most common set of programs used at this location. Some GIS software programs are:

- ARC/INFO (ESRI, Redlands, California)
- ATLAS/GIS (Strategic Mapping Inc., San Jose, California)
- GRASS (Public domain)
- INDRISI (Version 4.1, Clark University, Worcester, Massachusetts)
- LAS, LANDSAT ANALYSIS SYSTEM (EROS & NASA)

The cost of GIS software begins about $2000 and goes rapidly upward and requires workstations with extensive memory, digitizing equipment, and extensive operator training to begin work. While planned for the future, GIS software was not a quick and easy method for immediate display of graphical data. The alternative was desktop mapping software programs (Kawamoto, 1993) such as:

- ATLAS MAPMAKER (Strategic Mapping Inc., Santa Clara, California)
- ATLAS PRO (Strategic Mapping Inc., Santa Clara, California)
- GIV. GEOLOGIC INFORMATION VISUALIZATION (Russell Ambroziak, USGS, Office of Energy and Marine Geology; software now distributed in cooperation with Mary Washington College, Department of Environmental Science and Geology)
- HARVARD GEOGRAPHICS (Software Publishing Corporation, Santa Clara, California)
- MAPINFO (MapInfo Corp., Plano, Texas)
- MAPLINX FOR WINDOWS (MapLinx Corporation, Plano, Texas)

Desktop mappers generally cost between $400 and $1000 (Kawamoto, 1993). Besides cost of basic software, additional modules provide prepared maps showing U.S. county outlines, zipcode areas, transportation details, etc. These modules usually cost about $200 or more per module. Multi-user licenses cost much more.

Direct comparison of the various desktop mapping programs is not possible for most users because of cost and the time involved in learning to use each one. The choice becomes based on circumstances such as cost, local support, vendors' descriptions, what other users in an organization have used, etc. Some of the considerations used to compare products were:

1. Automatic display of data from standard relational databases such as dBase files linked to a map
2. Compatibility with imported and exported CAD, ASCII, Lotus, and Excel files
3. Clear display of data linked to map
4. Shading, ease of drawing, and color features
5. Easy to draw maps
6. Built-in spread sheets, graphs, data browsers.

Locating a user of any software product working on a project which paralleled a map index proved to be useless. The examples of projects done with desktop mapping software provided by the vendors or demonstrated by other users were of maps linked to simple short tables of data such as population, economic information, sample values, etc. The tables of data had short fields which were frequently just numbers. A geologic map index is a simple base map overlaid by outlines of many maps which frequently overlap. The tables of data for the California geologic map index needed to link to the display are made up of large fields (up to 220 spaces in the title field and 779 spaces for all fields) with a minimum of 3000 entries.

Another problem unique to the USGS library situation was that a database compiled in a federal facility should to be freely accessible to outsiders at all stages and ideally would not require purchasing expensive software or licenses to be used. For this reason a software package called GIV (Geologic Information Visualization) is being tried.

GIV is a package of mapping, image display and analysis software programs authored by Russell Ambroziak of the U.S. Geological Survey Office of Energy and Marine Geology and now distributed and supported by Mary Washington College, Department of Environmental Science and Geology. The programs use ASCII files converted to binary data. Spatial files including ARC/INFO files can be used to create maps. Maps can also be entered by digitizing or scanning methods. Color is used as an added dimension to data display. The software in this application allows for differentiating indexed maps according to scales by sorting the database into separate tables by scale and assigning various colors to a scale or range of scales.

GIV has been used to produce CD-ROM displays in the USGS's Digital Data Series. Explanations and examples of its use are available on the U.S. Geological Survey Open File 93-231 (Ambroziak and others, 1993).

Like any in-house program, the software does not come with handy manuals and is not available fr purchase in a cellophane wrapped box at a softv store. Authors of in-house programs will tend to tinker with the programming on a continuing basis and do not come out with periodic updates. The software may change over a period and there is usually no mechanism for informing a user that a "fix" has been made or that a program has evolved greatly over time. Help lines and 24-hour customer service are not available and the user is often relying on the goodwill of the programmer to get help when needed.

SUMMARY

The current method involves direct entry of bibliographical information into a Paradox relational database indexing by county, quadrangle, and coordinates and visually displayed on GIV software. A set of 450 citations from the Paradox database representing various types of publications has been used as the prototype database to see test how a proposed software such as the GIV programs will work. While the map index for California is being entered, the program can also be used for shorter data sets on California mapping such as maps in soil surveys published by the U.S. Soil Conservation Service and fault mapping and will eventually run on computers in the public area of the library.

REFERENCES

Ambroziak, R.A., Woodwell, G.R., Cook, C.A., and Wicks, A.E., 1993, Data, software, and applications for education and research in geology, Virginia: U.S. Geological Survey Open-file Report 93-231. [CD-ROM]

Kawamoto, W., 1993, Desktop mapping software: Computer Shopper, v. 13, p. 512-520.

Larsgaard, M.L., 1992, Accessing the world of digital spatial data: Western Association of Map Libraries Bulletin, v. 23, p. 188-192.

Blair, N.L., and Manson, C.J., 1994, Need for a national geologic mapping index database, *in* Wick, C., ed., Proceedings of the Geoscience Information Society, v. 24, p. 179-184.

Montgomery, B., 1994, Protecting data assets tops GIS migration concerns, GIS World, v. 7 no. 4, p. 38-40.

COOPERATIVE GEOLOGIC RESEARCH IN CANYONLANDS NATIONAL PARK, UTAH

J.L. Brown
R.F. Dubiel

U. S. Geological Survey
Denver, Colorado 80225

R.J. Schiller

National Park Service
Denver, Colorado 80225

Abstract -- Canyonlands National Park, Utah, is located on the Colorado Plateau, an elevated structural platform comprising an area of 150,000 square miles. The structure within the Colorado Plateau is relatively simple: the mantle of sedimentary rocks is flat lying, folded by broad monoclines oriented north-northwest that are occasionally broken by faults. The park includes a portion of the Paradox Basin, a major northwest-southeast trending structural depression that formed during Middle Pennsylvanian time in association with uplift of the adjacent ancestral Rocky Mountains. The area exposes rocks from mid-Paleozoic to Cretaceous in age, but the Permian-Triassic interval is regarded as especially unusual because it represents an extreme paleoclimate that involved the global deposition of more red beds and evaporites than at any other geologic time. The study area contains large oil and gas reserves, unique species and habitat resources, and spectacular geologic exposures of exceptional scenic and recreational value. New models and principles have recently been developed in sedimentology, sequence stratigraphy, and global plate tectonics. These models and techniques coupled with the acquisition and analyses of new data will improve our understanding of unresolved problems of Colorado Plateau/Canyonlands geology such as timing and extent of episodic transgressive and regressive events, ultimate causes of regional tectonism, and refinement of the paleogeography of the Permian-Triassic interval. Through a joint funding agreement between the U.S. Geological Survey (USGS) and the National Park Service (NPS), the project was initiated in June of 1994 and is scheduled for completion in December 1997. Because the USGS has ongoing geological research projects in the area, our objectives include combining resources to solve geologic and resource management problems not only for the park, but for local, state, and national data users. The new USGS National Spatial Data Infrastructure (NSDI) initiative has critical need for new geologic framework data sets in digital format through Geographical Information Service (GIS) technology. Objectives include the development of procedures, standards, and data acquisition for base maps and GIS analysis. Results provide a basis for continued scientific research and assist in land planning and management tasks. The project will deliver a large-scale geologic survey of the park, a national shared GIS data base, and research reports on topical studies. Ongoing geologic research provides greater detail on certain critical geologic relationships, exploration for additional geologic structures, and relating local deformational features to a regional scheme.

INTRODUCTION

Canyonlands National Park (CNP) in southeastern Utah was established in 1964 to preserve an immense untrammeled wilderness of fascinating geology (figure 1.) The park encompasses 527 square miles (1365 km^2) in the heart of the Colorado Plateau geologic province and has been intricately dissected by literally hundreds of canyons. Park headquarters is located in Moab, Utah, about 50 miles (80.5 km) southeast of Green River, Utah, and about 51 miles (82.1 km) west of Grand Junction, Colorado. The park is drained by the Colorado and Green Rivers, whose confluence is an important scenic feature for visitors.

The park is divided into three administrative and interpretive Districts: Island in the Sky, The Maze, and The Needles. Each District preserves a vast tract of different geological formations that are distinguishable from those in other Districts by

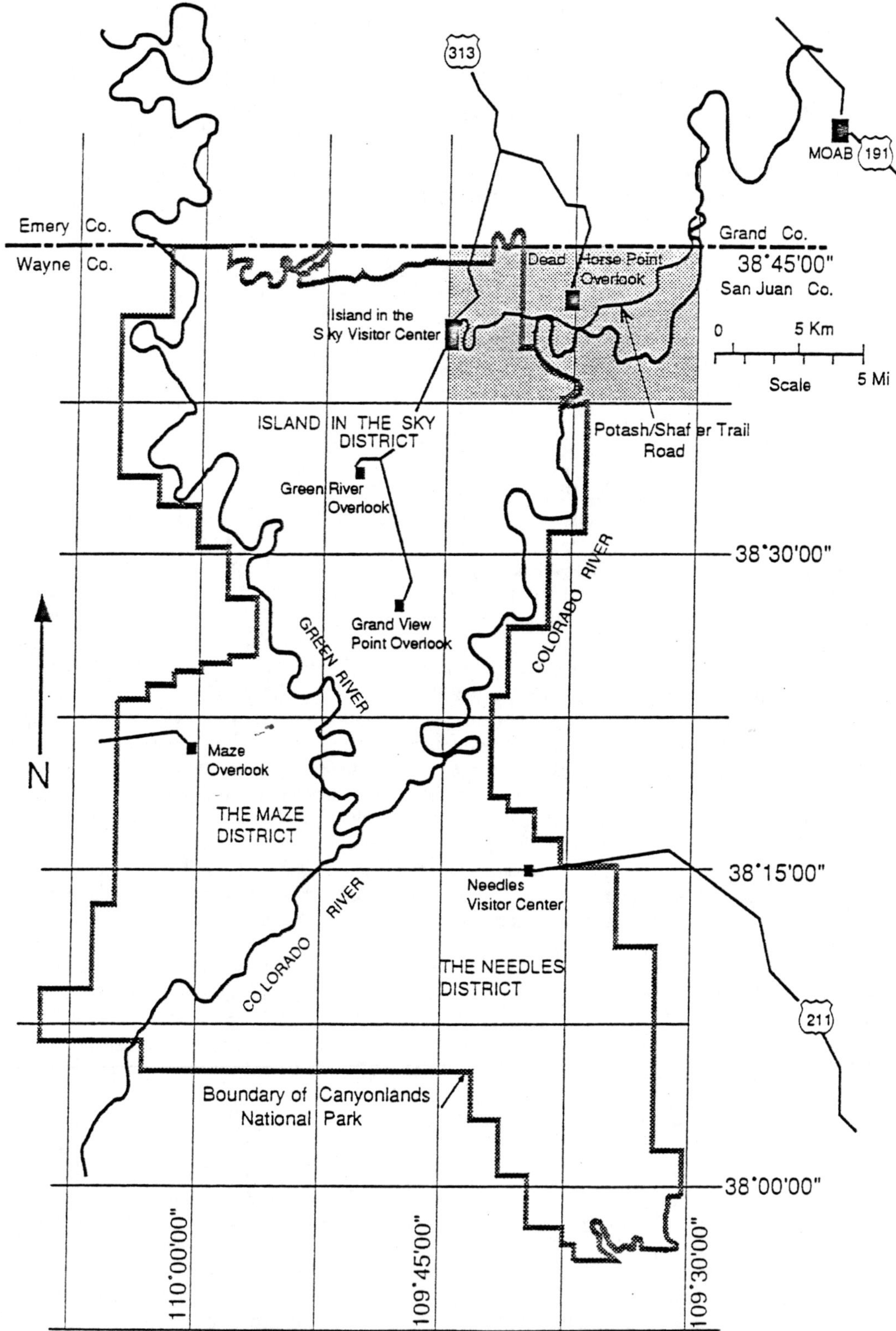

Figure 1. Location map of Canyonlands National Park, Utah. Shafer Basin and Musselman Arch quadrangles are shaded.

special styles of physiographic expression. The physiographic expressions relate to important geological processes relevant to the geological evolution of the park. Physical features such as resistance to erosion, thickness of strata, and linear features having major structural significance shape the landforms and tell a unique story in each district. An east-west cross-section across the park at about 38° 45' latitude shows three prominent topographic levels or "Three Worlds." The highest level is appropriately named "Island in the Sky." It occupies the highest point between the Colorado and Green Rivers. The second level is located half-way down the canyon walls where prominent flat benches and platforms are dissected by vertical cliff-bordered canyons. The lowest level of all may be seen only from certain vantage points -- the channels and flood plains of the Green and Colorado Rivers.

The Colorado Plateau is an elevated structural platform ranging in altitude from 5,000 to 7,000 feet (1524 to 2134 m) and comprising an area of 150,000 square miles (388,500 km2). The structure within the Colorado Plateau is relatively simple: the sedimentary rocks are flat lying and in places are folded by broad monoclines oriented north-northwest that are occasionally broken by faults.

The park includes a portion of the Paradox Basin, a major northwest - southeast trending structural depression that formed during Middle Pennsylvanian time in association with uplift of the adjacent ancestral Rocky Mountains. The park exposes rocks from mid-Paleozoic to Cretaceous in age. The Permian-Triassic rocks are of special interest because they record a paleoclimate of extreme heat and aridity that resulted in the global deposition of more red beds and evaporites than at any other geologic time. The Paradox Basin contains large oil and gas reserves, uranium resources, and tar sands. In Canyonlands National Park, the arid desert climate serves as habitat for a unique association of plant and animal species.

PURPOSE OF COOPERATIVE GEOLOGIC RESEARCH

Our purpose is to provide a geologic survey of Canyonlands and a Geographic Information System (GIS) database to be shared by the U.S. Geological Survey (USGS) and National Park Service (NPS), and made available to local, state, and national data users. Anticipated products include USGS Open-file geologic maps, USGS *Geologic Quadrangle* (GQ) maps, USGS *Bulletins* on the interpretive geology of the park, topical studies, and associated reports.

The GIS data set housed in the National Geospatial Data Clearinghouse provides a powerful and cost-effective tool to develop responses to preservation and management issues, to answer requests for public information, and to provide a basis for continued scientific research in the region. The project addresses the Department of Interior's national mandate to authorize, encourage, and implement interagency agreements for mutual benefit and cooperation, and it draws upon previous collaborations (Brown and Davila, 1993; Brown and others, 1993; Brown and Davila, in press; Brown and others, 1994). The general management plan for Canyonlands National Park provides for the protection of historic, geologic, and cultural resources; management, monitoring, and restoration of habitat; and identification of visitor needs, expectations, and impacts. Park interpretative programs provide sound science and environmental education and allow visitors to explore the value of parks as laboratories for natural resource studies.

The collaboration provides large-scale geologic mapping, GIS data sets, and results of other geologic research to the park. The results help NPS develop responses to preservation and management issues and provide a basis for continued scientific research in the region. The objectives of new mapping include providing new data on the Paleozoic, Mesozoic and Cenozoic depositional history of the region; greater detail on certain critical stratigraphic relations; assessment of possible mineral and energy resources; and relating local fundamental deformational features to a regional scheme.

Early small-scale mapping was done prior to the development of modern basin analyses concepts, and before new subsurface structural and paleo-geographic analyses were available, so that details of correlation, stratigraphic succession and extent, and origin of geologic units were not well established. Topics of continuing study in Colorado Plateau geology include the tectonic evolution of the Paradox Basin, origin of a prominent northeast trending fracture zone, origin of Upheaval dome (a salt dome or astrobleme), origin of graben swarms in the Needles District, and origin and timing of emplacement of bounding laccolithic groups such as the Abajo Mountains.

VISUALIZING THE LANDSCAPE

Development of a shared GIS includes establishment of procedures, standards, and data acquisition for base maps. After development, the data bases are catalogued and archived with the

USGS National Geospatial Data Clearinghouse which disseminates the new geologic and topographic data sets to all users. Because GIS databases can have different vertical resolution and topographic contour detail, an intrinsic feature of merging GIS data bases requires the realistic integration and portrayal of topographic form and geologic overlay.

The strategy for sharing geoscience and resource management GIS data requires integrating USGS and NPS spatial structuring (raster versus vector). Selection of type of spatial structuring is designed to provide different kinds of information and analyses. Resource and inventory data generated by NPS parks, is set up in a two-dimensional raster-based GIS in the Geographical Resources Analysis Support System (GRASS) data model. GRASS, established in the early 1960's, was the first system available to users for scientific and modeling applications in a workstation platform.

In contrast, the USGS, in order to provide topographic bases and geologic modeling, established a three-dimensional vector-based GIS in the ARC/INFO system. The ARC/INFO data model is complex and is designed to analyze two features of geography: features as they are oriented in space and features associated with a theme, e.g., geology.

GRASS is very effective in resource management applications where remote sensing input is significant and the concept of continuous geographical variation (such as in biological systems) is more acceptable. Raster utilizes a simple data model and cheap technology, and provides easy data collection and processing. Raster requires no topological (three-dimensional) processing and only limited handling of themes. Raster stores, processes, and displays graphic data as images that present information as values in uniform grid cells or pixels. Remote sensing satellite systems scan, capture, and transmit data initially in raster form. The topographic data base used as underlay for resource management information is formulated using digital line graph (DLG) form, which is planimetric, contour-based, and two-dimensional.

The ARC/INFO system, which is three-dimensional and vector-based, was launched in 1982 by Environmental Systems Research Institute (ESRI) a non-profit research organization established in 1970. In the name "ARC/INFO", "ARC" refers to the topological (three-dimensional) data structure and algorithms, "INFO" to the tabular data structure (e.g. rock strata) and algorithms, and "ARC/INFO" to the composite data model and associated processes. Vector data are represented by strings of coordinates representing the true position of features on Earth and are shown by tics (points), arcs (lines), nodes (end points of arcs), and polygons (areas). A vector based GIS manipulates or displays polygons by their attributes. Vector data are matched with topography using a digital elevation model (DEM) also called a digital terrain model (DTM) in which elevation data are three-dimensional. ARC/INFO visualization is in either perspective, or "2.5-dimension," or in three-dimensions. "2.5-dimension" visualization refers to aerial oblique views of terrain in which the azimuth direction and elevation of a view can be selected and displayed. ARC/INFO speed of data analyses is slow because of the complex data model, difficult and complex data collection and processing, and associated topologic analysis. In contrast, where functions such as presence, absence, or overlap of a feature is required, the speed of analysis in the raster data model is fast.

Two-dimensional raster function does not provide the techniques of spatial data handling required by geoscientists in conceptual data modeling. The great advantage for geoscientists in vector modeling is that many existing geologic questions can be more effectively answered in three-dimensional modeling, completely new questions can be posed, and this form of analysis can be continuously extended. Vector modeling produces models that can be quickly made and improved and enhances a geologist's ability to communicate with his or her peers and the public.

Previous practice in geoscience has been to present the results of a particular study or model as a paper map. However, utilizing digital data bases, models can be continuously replaced by new models, each of which may be capable of many visualizations. The geologic feature itself is usually defined by sampling or selection of parameters established in the data model. Only discrete geologic features, such as a perched aquifer, salt dome, or fault bound block (horst), could be represented well in raster format. Geologic features that represent variable density, transient phenomena, or part of a continuum such as a sedimentary facies are best analyzed in vector systems.

Methods of sedimentary geology in basin analysis utilize diagrams of vertical profiles generated from either subsurface well-log analysis or measured sections of strata. Once digitized, data, even from a sparse data set, can be analyzed from selected perspectives in time and space and applied to a basin analysis model or compared to other geological features or models. A time slice analysis is possible along a selected geological stratum for the generation of paleogeographic maps. Geological data constraints can also be identified: (1) geological discontinuities (e.g., fault types), (2) regional structure (e.g., dip/strike and fold patterns), (3) sedimentation patterns (e.g., alluvial fan systems and delta systems),(4) volcanic or plutonic history (shape, depth, emplacement history), (5) processes within a sedimentary environment (e.g., glacial weathering). Vector analyses of geological features reconstructs the structural and sedimentologic evolution of basins, or any system where the data model changes rapidly.

METHODS

Previous geologic mapping in Canyonlands National Park was conducted from 1977 to 1980 and was published in 1982 as a small-scale (1:62,500) regional map (Huntoon and others, 1982). Older photogeologic mapping was published by USGS between 1953 and 1955 as large-scale (1:24,000) geologic maps, but they are now out of print. The *Geologic map of Canyonlands National Park and vicinity, Utah* (Huntoon and others, 1982) is digitized in a GRASS GIS data base.

Collaboration between USGS and NPS resulted in the purchase of the 29 large-scale (1:24,000) quadrangles and the digital elevation models of the topography that were needed for the entire park. The preparation of new geologic maps are prioritized according to geologic complexity and visitor use. Two quadrangles, Shafer Basin and Muscleman Arch, have been selected for initial field work.

Published and unpublished geologic data were compiled onto registered velum topographic base maps for the two quadrangles. Stable base velum is the required base material for GIS digitizing and further map publishing processes. In addition, new data were compiled from recent aerial photography of the park using the Kern computerized photogrammetric mapper (PG-2). The compilations serve as working drafts for the actual on-the-ground field work and are finalized after field mapping is completed.

The geologic map is first published as an author-prepared, black and white, USGS Open-file map and report, which is quickly available to the NPS and the geologic community. The completed product is a full-color USGS GQ map. After the completed geology is drafted onto the base map, a clear-film blackline copy is made and then scanned on a digital Drum Scanner and the data converted into ARC/INFO digital coverage. The ARC/INFO digital coverage is edited to eliminate any errors and is then translated into a GRASS data model. If further analysis is needed, such as three-dimensional visualization or perspective graphics, the ARC/INFO digital coverage is manipulated by a Triangulated Irregular Network (TIN) program that matches the geology to the ground and facilities edge mapping and compositing of completed quadrangles into a geologic map representing the whole park.

CONCLUSIONS

This project provides a geologic survey of Canyonlands National Park and a GIS data base shared by the U.S. Geological Survey and the National Park Service, that is available to local, state, and national data users. The GIS data set provides a powerful and cost-effective tool to develop responses to preservation and management issues, to answer requests for public information, and to provide a basis for continued scientific research in the region.

REFERENCES

Brown, J.L. and Davila, V. Jr., 1993, Great Basin National Park and the U.S Geological Survey Cooperate on a geologic mapping program: National Park Science Bulletin, v.13, no. 2, p. 6-8.

Brown, J.L., Miller, E.L., Miller, D.M., Crane, M.P., and McCarthy, P.T., 1993, Geographic information systems (GIS), mapping, and geology in Great Basin National Park, Nevada: Cooperative interagency research: Geological Society of America Abstracts with Programs, v. 25, no. 5, p. 14.

Brown, J.L., and Davila, V. Jr., in press, Interagency geologic research and geographic information systems (GIS) analyses in Great Basin National Park, *in* van Riper III., C., ed., Proceedings of the Second Biennial Conference on Research in Colorado Plateau National Parks, Flagstaff, AZ October 23-28, 1993: National Park Service Transactions and Proceedings Series; NPS/NRNAU/NRTP, 1995, no. 11. p. 20

Brown, J.L., Dubiel, R.F., and Schiller, R.J., 1994, Cooperative geologic research in Canyonlands National Park, Utah: Geological Society of America Abstracts with Programs, v. 26, no. 7, p. 65.

Huntoon, P.W., Billingsley, G.H., and Breed, W.J.,

1982, Geologic map of Canyonlands National Park and vicinity, Utah: The Canyonlands Natural History Association, Moab, Utah, 1 sheet, scale 1:62,500.

THE GLOBAL CHANGE DATA AND INFORMATION SYSTEM (GCDIS):
THE ROLE OF LIBRARIES IN
DEVELOPMENT AND IMPLEMENTATION

Gordon S. Banholzer, Jr.

Library Services Branch
NASA/Goddard Space Flight Center
Greenbelt, Maryland   20771

Abstract – The United States government has initiated a comprehensive scientific research program to investigate Earth processes as an integrated system, including potential long term changes in the global environment. Participating federal agencies are developing programs and systems to simplify the task of accessing, obtaining and using the data generated by this research as well as existing information resources. The library community has established a role in providing input and expertise on the design, evaluation, and promotion of the Global Change Data and Information System (GCDIS).

INTRODUCTION

Recently, the library community has played a proactive role in making contributions to the development of programs and systems being created to enable access to global environmental and earth sciences information. This effort has helped establish connections and interchange of ideas between the library community and data managers primarily responsible for the development and implementation of the programs and systems.

GLOBAL CHANGE RESEARCH

In 1989 the United States government formally established a policy for conducting a major series of coordinated research efforts aimed at understanding the scope and rate of natural and human-induced global environmental change (Committee on Earth Sciences, 1989, 1990; Committee on Global Change, 1990). The United States Global Change Research Plan (USGCRP) is being conducted as part of a series of international research programs that address various components of environmental change (e.g. International Geosphere-Biosphere Programme). There have been a series of research programs and reports since the mid-1970s directed towards the understanding of global environmental change (e.g. Global Atmospheric Research Program). None, however, compare in scale to the USGCRP and associated international efforts.

An array of U.S. Government agencies and organizations have responsibilities in the USGCRP. These include the National Aeronautics and Space Administration (NASA), National Oceanic and Atmospheric Administration, United States Geological Survey, Environmental Protection Agency, Department of Defense, Department of Energy, National Science Foundation, Department of Agriculture, Smithsonian Institution, Tennessee Valley Authority, Office of Management and Budget, Office of Science and Technology Policy, Department of State, and Department of Health and Human Services.

GLOBAL CHANGE DATA AND INFORMATION SYSTEM

The USGCRP is expected to generate massive amounts of data from earth-orbiting satellites as well as less voluminous data from ground-based validation and verification studies. Digital data volumes will be measured in hundreds of terabytes (trillions of bytes) per year. The principal space-based component of the research program is the Earth Observation System being planned and developed by NASA (Asrar and Dozier, 1994)

In order to contend with these data in a more effective and efficient manner, an overall strategy for computer-based data and information systems was developed (NASA, 1992). It was recognized early in the planning for the USGCRP that primary research scientists would need to have satellite data processed and compiled into useful data sets in a timely manner. The data products would have to be accessed and obtained via telecommunication networks rather than as computer tapes mailed from the data centers because of their sheer volume and variety.

A committee, the Interagency Working Group on Data Management for Global Change (IWGDMGC) was established in the late 1980's to coordinate the planning and implementation of a data and information system by the federal organizations involved. The IWGDMGC grew to include nearly 20 cabinet level departments as well as independent agencies and executive branch offices.

A Global Change Data and Information System (GCDIS) was conceived as a program under which access to data and information by the participating agencies would be facilitated. The system uses computer and telecommunications systems for the core technological infrastructure to allow ready access to digital data sets, metadata (textual descriptive information about the digital data sets), and other information resources. However, the GCDIS should not be viewed as just the computer systems; it encompasses the entire effort to link relevant data and information and make it available to the user.

Each agency participating in the GCDIS has the responsibility for developing its own system and implementing it at its own pace. A key requirement, however, is that each system be interoperable with the other systems. Implementation plans for all of the individual agencies are being finalized in 1995. NASA's portion of the GCDIS is the Earth Observation Systems Data and Information System (EOSDIS). Version 0 of EOSDIS was made available via the Internet and the World Wide Web in April 1994. Version 1 is planned for release in 1997 and Version 2 in 1999.

ROLE OF LIBRARIES

Information relevant to global environmental change will not exclusively be digital data sets generated by satellite-borne sensors. Because there is a strong historical component to the USGCRP - researchers need to have access to and understand past measurements of the atmosphere, the oceans, the solid earth, and the sun - data sets and text resources that reside in libraries will have significant importance. Much of this information is in print or microform formats that may not be digitized in the foreseeable future, if ever.

In 1990 a number of librarians and other information professionals in the Washington DC area formed an ad hoc group to discuss issues raised by the expected significant increase in global change research funding. The group began laying the foundation for an enhanced role for the library community in the planning and development of the various data and information systems.

One of the primary goals was to find avenues for providing input to the data managers and to influence the development of the data and information systems.

Contact was established with the IWGDMGC, which was organized as a group of principal representatives from each of the agencies involved, with several subgroups created to deal with functional areas such as the content requirements, interoperability of the systems developed by different agencies, etc. The ad hoc group began an effort to petition the IWGDMGC for formal recognition as a subgroup. After a number of meetings and presentations, the group was officially established in 1992 as the Library Information Subgroup (LIS). The two other extant subgroups, after a reorganization by the IWGDMGC, are Contents and Access.

The LIS played a valuable role in the period 1990-1992 in raising issues of broader access to the GCDIS to the IWGDMGC. It is clear from documents published in 1989 and 1990 by the organizations responsible for the USGCRP that the planned data and information systems were intended to be composed of data and information for primary research scientists. It was expected that the systems would exist to provide them with easier access to comprehensive data sets. There was little indication of any recognition or belief that a broader community of policy makers, planners, educators, and students would need or want access to global change information.

IWGDMGC aware of the need for serving a broader community, the LIS currently provides the library/ information center perspective and expertise for the design, evaluation, and promotion of the GCDIS. More specifically, the LIS's role in the IWGDMGC can be outlined as follows (LIS, 1994):

1. Building an infrastructure of libraries and librarians for GCDIS implementation
2. Evaluating GCDIS from a library user's perspective and providing user needs analysis
3. Linking data resources to information resources for knowledge management
4. Promoting GCDIS to libraries and developing approaches to user education
5. Advising on data and information processing standards and systems from the library perspective.

During the last two years, the LIS has facilitated several projects related to information search and retrieval, a key element of the GCDIS (Rand, 1995). If persons other than primary research scientists are to have ready access to the information resources of the GCDIS, user interfaces and search methods need to be clear and effective. New interfaces such as the World Wide Web and the variety of browser software written for it should be evaluated. In addition, methodologies beyond Boolean should be investigated in order to provide state-of-the art capabilities to the system.

First, members of the group, as well as members of other IWGDMGC subgroups conducted evaluation of a natural language, semantic network retrieval software package developed commercially by Conquest Software. The evaluation involved a variety of databases (bibliographic, image, and data sets) contributed by several IWGDMGC participants.

Second, the Assisted Search for Knowledge (ASK) project involves researchers from the National Agricultural Library and employs the Conquest software, a WAIS search engine, and a graphical user interface written for multiple platforms in order to test and demonstrate a system that will allow users with a range of skill levels to access and retrieve data and information from the GCDIS.

Third, the LIS is involved with a study of general user access to the GCDIS. The Library Access Search and Retrieval (LASR) study is being conducted by a consortium of university/college libraries, public schools, a state museum, and an array of environmental groups in the state of Virginia. Funding is provided by NASA's Mission to Planet Earth program office, with oversight by the Universities Space Research Association (USRA).

Results of demonstration projects such as ASK and LASR will provide valuable information on the ways in which users ranging from researchers to the general public interact with the systems being developed. Further evaluations will very likely be required as the GCDIS develops. It is the intent of the LIS to involve interested groups and individuals in such projects. In some cases the projects may be funded; in other cases, the evaluations may require volunteers.

CONCLUSION

The establishment of the Library Information Subgroup has brought the expertise and views of the library community to the development and implementation of the Global Change Data and Information System. As this program expands in the coming decade, opportunities for increased participation and input will exist for librarians and other information professionals to influence its development.

SOURCES AND CONTACTS FOR GLOBAL CHANGE INFORMATION

Global Change Research Information Office
User Services
2250 Pierce Road
University Center, Michigan 48710
(517) 797-2730
(517) 797-2622 fax

e-mail   help@gcrio.org
telnet   gopher.gcrio.org login as "gopher"
gopher   gopher.gcrio.org

Global Change Data and Information System

gopher   gopher.gcdis.usgcrp.gov
          esdim2.nodc.noaa.gov
WWW   http://www.gcdis.usgcrp.gov
          (under construction

Global Change Master Directory

WWW   http://gcmd.gsfc.nasa.gov

EOSDIS Version 0

WWW   http://gcmd.gsfc.nasa.gov/-gcmdeos.html

Library Information Subgroup

email   gcdis@usgcrp
gopher   gcdis.lib@earth.usgcrp.gov
          LIS information for the GCDIS
          (gopher under construction)

REFERENCES CITED

Asrar, G., and. Dozier, J., 1994, EOS: science strategy for the earth observing system: Woodbury, New York, AIP Press, 119 p.

Committee on Earth and Environmental Sciences, 1992, The U. S. global change data and information program plan: Washington, D.C., Government Printing Office, 94 p.

Committee on Earth Sciences, 1989, Our changing planet: A research strategy for global change research: Washington, D.C., Government Printing Office, XXX p.

Committee on Earth Sciences, 1990, Our changing planet: The FY 1990 research plan: Washington, D.C., Government Printing Office, 184 p.

Committee on Global Change, 1990, Research strategies for the U.S. global change research program: Washington, D.C., Government Printing Office, 291 p.

Library Information Subgroup, 1994, Submission for Global Change Data and Information System implementation plan (draft).

National Aeronautics and Space Administration, 1992, EOS data and information system (EOSDIS): Washington, D.C., Government Printing Office, 31 p.

Rand, R., 1995, Library Hi-Tech, (in press)

THE MAGELLAN DATA PRODUCT ACCESS SYSTEM

Edward A. Guinness
Susan Slavney
Thomas C. Stein

Department of Earth and Planetary Sciences
McDonnell Center for the Space Sciences
Washington University
One Brookings Drive
St. Louis, Missouri 63130

Abstract -- The Magellan Mission to Venus returned over 400 gigabytes (400 x 109 bytes) of digital data. The data products consist of radar images and altimetry, radiometry, and gravity data in the form of orbital tracks and global map images. These data products are stored on a combination of CD-ROM and CD-WO media. The choice of media is governed by factors such as dataset size, complexity, andanticipated rate of use. The Planetary Data System Geosciences Node has developed a system toprovide long-term access to the Magellan data. An on-line catalog allows selection of subsets of data by criteria such as product name, location, time of acquisition, and geographic feature. Image products selected by searching the catalog can be displayed on the user's terminal screen. The system has a dataorder mechanism so that scientists can request copies of data products. Data is distributed on CD-ROM,magnetic tape, CD-WO, or by electronic transfer over the Internet. A series of CD jukeboxes allows dataarchived on CD media to be near on-line and made available for electronic transfer. The ClementineMission to the moon and, in the near future, the Mars Global Surveyor mission will provide data archives of comparable size to the Magellan archives. These future planetary data archives willpresent new challenges for providing data access for scientists, particularly in terms of electronic data distribution.

INTRODUCTION

Scientists interested in image data collected by missions to the planets were once limited to using photographic prints made from the digital data collected by spacecraft sensors. Those who wanted to use the data digitally needed large, expensive computers. In the past decade or two this scenario has become obsolete; computers became smaller and cheaper just as planetary missions began to return increasingly large amounts of data. Most recently, the Magellan Mission to Venus produced over 400 gigabytes (1 gigabyte = 109 bytes) of data, more data than all previous planetary missions combined, making traditional methods of distributing data to scientists unworkable.

The problem of providing scientists with long-term access to Magellan data products can be viewed as having three main issues. The first issue is to archive the data and documentation on a stable media to ensure that the data can be accessed over a long period of time. The second issue is to provide scientists with tools to navigate through the very large datasets to locate subsets of data that are suitable to their research projects and that are of manageable size. The third issue is to provide an efficient mechanism for distributing selected data products to the science community. To address these issues, the Geosciences Node of NASA's Planetary Data System (PDS) (Arvidson and Dueck, 1994) has developed a system to provide access to Magellan datasets. The Magellan data access system consists of an electronic catalog component for searching through and selecting subsets of datasets and a data order function that enables scientists to request copies of data products. The catalog and ordering components are available via the Internet using a remote login or via the World-Wide Web using a browser such as Mosaic (using the URL "http://wwwpds.wustl.edu/"). The system supports several different data distribution methods that include shipment of data on CD-ROMs, magnetic tape, and CD write once (CD-WO) disks and electronic data transfer over the Internet.

In this paper the Magellan data access system is characterized. A brief review of the Magellan Mission and the major datasets produced by the mission is first given. The method of archiving the Magellan datasets is then described, concentrating on explaining the rationale for choosing a particular archive media. The system catalog structure, search mechanisms, and the data order function are discussed. This is followed by a explanation of the electronic data distribution operation. Finally, issues related to providing access to future large digital planetary data archives are considered.

MAGELLAN MISSION AND DATA PRODUCTS

The Magellan Mission to Venus was designed to provide global views of the planet in order to study the geologic processes that have operated on Venus (Saunders and others, 1992). The Magellan spacecraft was launched in May 1989. It began mapping the planet in September 1990 after being inserted into orbit around Venus. Scientific observations continued until October 1994 when the spacecraft entered the planet's atmosphere and communications with the spacecraft were lost. While the spacecraft was operating, it collected data that covered about 98% of the planet's surface. As a result, the mission was able to produce global datasets of radar backscatter characteristics, topography, microwave emissivity, and gravity (Saunders and others, 1992).

The Magellan spacecraft carried a radar system that produced synthetic aperture radar (SAR) images from orbit when operated in an active mode (Saunders and others, 1990). The same system was also used in a passive mode to measure the amount of ambient microwave energy emitted from the planet's surface (Pettengill and others, 1992). The spacecraft had a second microwave antenna that was used to measure topography and radar scattering properties (Saunders and others, 1990; Ford and Pettengill, 1992). In addition, precise tracking of the spacecraft position and changes in its orbit yielded information on the planet's gravity field (Sjogren and Konopliv, 1994). A number of datasets have already been generated from the data returned by these instruments. Additional datasets continue to be produced by further processing of the data. The Magellan datasets that are currently part of the Magellan data access system are briefly described in the following paragraphs.

The SAR image data acquired on each orbit are formatted into a data product called the Full- resolution Basic Image Data Record (F-BIDR). Figure 1 shows an example of a small portion of two F- BIDRs acquired on consecutive orbits. F-BIDRs are processed to have 75 meter pixel spacings and are about 300 pixels wide. However, each F-BIDR product contains data for an entire orbit, and data collection during one orbit could go nearly from pole to pole. Thus, F-BIDRs can be very long and contain as many as 220,000 lines. The aspect ratio of F-BIDRs can best be described as a long, thin

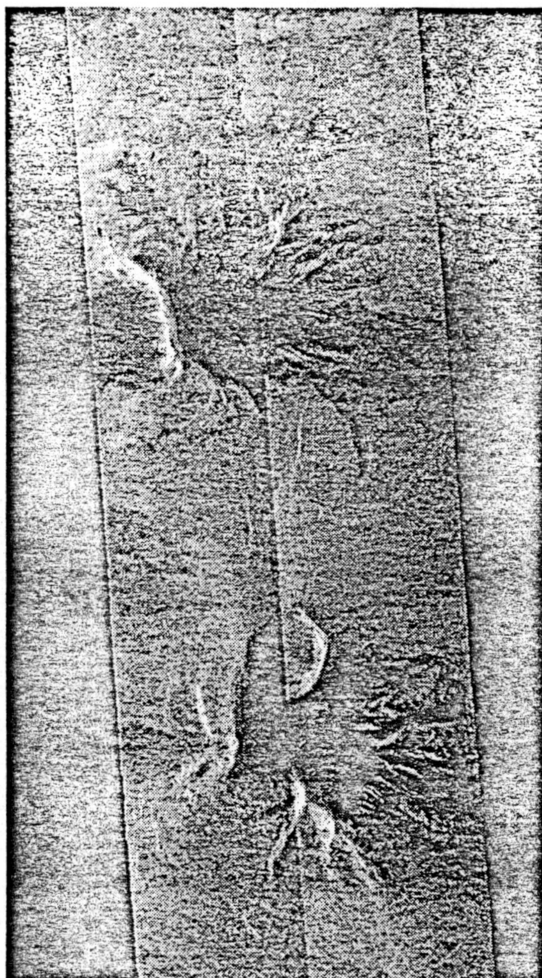

Figure 1. A small portion of two Magellan F-BIDRs is shown. The F-BIDRs show the summit of the volcano seen in Figure 2. Each F-BIDR strip is about 20 km wide. A complete F-BIDR is about 100 times longer than what is seen here.

strip since they can cover an area as long as about 18,000 km and only about 20 km wide. The SAR image data in an F-BIDR has been projected as a sinusoidal equal-area map projection. Stored along with the image data are ancillary files that contain processing and instrument engineering information (Saunders and others, 1992). Thus, each F-BIDR product can be up to 140 Mbytes in size. The Magellan archive contains about 6000 F-BIDR products, which provide nearly global coverage (98%) of the venusian surface. The total F-BIDR dataset is over 300 gigabytes in size.

The unusual aspect ratio of an F-BIDR makes the product somewhat difficult to work with as an image.

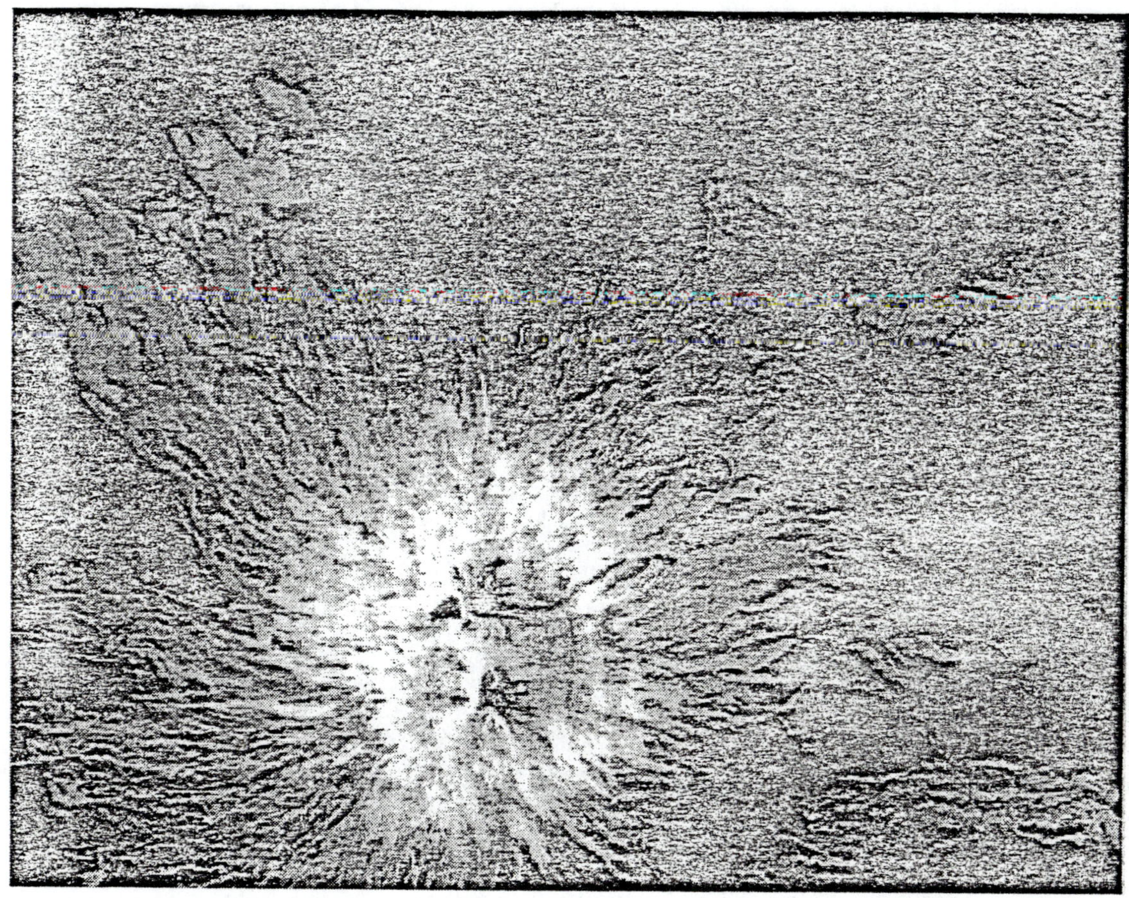

Figure 2. This example of a Magellan F-MIDR product is centered on a volcano located in the equatorial region. The frame is about 540 km from top to bottom. The vertical stripes are due to slight orbit to orbit variations in the F-BIDRs used in the mosaic.

As a result, digital map products were created by the Magellan Project that combine several F-BIDR products into a more traditional image format. These products are referred to as MIDRs for Mosaicked Image Data Record. There are currently full resolution (F-MIDR) and reduced resolution (C-MIDR, where the C stands for compressed) mosaics. The size of a MIDR product is 7168 lines by 8192 samples, or nearly 58 Mbytes. F-MIDRs have the same resolution as the F-BIDR products, that is, 75 meter pixel spacing. C-MIDRs are produced by averaging together groups of pixels from the full resolution data. Thus, C-MIDRs cover a larger area than F-MIDRs. There are three varieties of C-MIDRs based on how much the original resolution has been reduced. The three types of C-MIDRs have resolutions reduced by factors of 3, 9, and 27 that produce pixel spacings of 225, 675, and 2025 meters, respectively (Saunders and others 1992). Figure 2 shows an example of an F-MIDR, whereas Figure 3 is an

example of a C-MIDR that includes the region seen in Figure 2. Because of Magellan Project cost restraints, F-MIDRs were produced for only about 20% of VenusU surface. However, the coverage of C- MIDRs is nearly global. There are 1255 distinct MIDR products in the Magellan archive, which make the total MIDR dataset about 73 gigabytes in size.

Additional datasets available in the Magellan data access system include altimetry measurements, radiometry data in the form of passive microwave emissions, estimates of meter-scale surface roughness, and spacecraft accelerations due to irregularities in the Venus gravity field. These datasets exist in two formats. In the first format, known as a time-series format, the data are ordered by time of data acquisition. The second format is a global map stored as a digital image. In the time- series format the altimetry data have resolutions between about 10 and 30 km and the radiometry data have resolutions of 10 to 90 km, with the resolution in

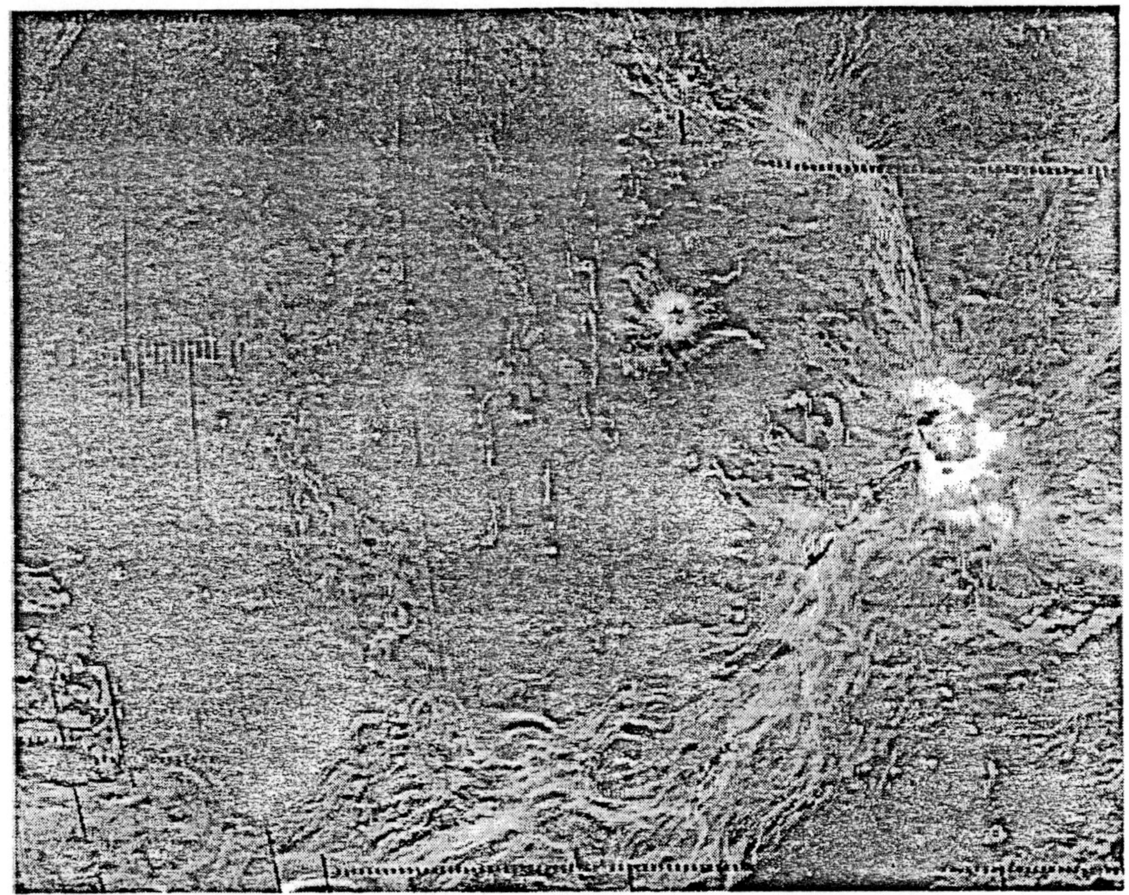

Figure 3. This example of a Magellan C-MIDR product has been compressed by a factor of 9 relative to an F-MIDR. It shows the region around the volcano seen in Figure 2 (the bright area in the upper third of the image and just right of center). Black stripes and patches are due to missing data. The frame is about 4800 km from top to bottom.

both cases being dependent on the spacecraft altitude (Pettengill and others, 1991). Gravity measurements were acquired about every two seconds, with the spacing along the spacecraft orbit also dependent on the spacecraft altitude. Global map images currently exist for the topographic and microwave emission data. These maps were produced with a 5 km pixel spacing (Pettengill and others, 1991; Ford and Pettengill, 1992). Global maps for the gravity data are still under construction. The total size of the altimetry, radiometry, and gravity datasets currently available through the Magellan data access system is about 15 gigabytes.

ARCHIVING MAGELLAN PRODUCTS

As described in the introduction of this paper, one of the issues in providing scientists with long term access to Magellan data is how to archive the products. The Magellan data products have been produced using standards established by the NASA Planetary Data System (PDS) for archival data products (Arvidson and Dueck, 1994). These standards are meant to ensure that data in an archive are properly validated for scientific content and that the documentation and ancillary information needed to work with the data are available. A detailed discussion of the PDS archiving standards is beyond the scope of this paper. Instead, we will focus on explaining how the choice of archiving media was made for the Magellan data products.

All of the data products available in the Magellan data access system are archived on CD media, either on CD-ROMs produced in a factory or on write-once CDs produced by the PDS. CD-ROM and CD-WO media were chosen because they provide high density storage (up to 680 Mbytes within a small space), are durable, are likely to be stable over long periods of time, and are easy to replicate for distribution. The use of CDs avoids many of the problems associated with magnetic tape media, such as damage from environmental factors like magnetic fields and dust, or from stretching and tearing due to repeated use. The length of time that a CD will last is uncertain. Claims from the CD industry that the disks will last many 10's of years are hard to test until CDs have been in use for a long period of time. However, we are not aware of any problems with Magellan CDs in the 4 to 5 years of their use.

The issue relevant to archiving Magellan data products was the choice between CD-ROM and CD-WO media. The choice was dependent on a combination of factors, including rate of use for a dataset, dataset size, and cost of production for the two different formats. The number of people wanting to use a dataset will determine the number of copies needed. The best way to satisfy a large number of users requesting a dataset is to make a copy for each user. This is easy to do with CD-ROMs that are mass produced by a vendor. On the other hand, making a large number of copies of a CD-WO volume would be very time consuming and expensive. The size of a dataset will determine the number of volumes needed to archive that dataset, and thus, have a direct impact on the cost of producing the archive. In addition, for a very large dataset, such as the F-BIDR dataset, users may not want to store a large number of volumes in their own facility, particularly if they only need to access a subset of the data.

The cost and effort to prepare data for archive volumes is generally the same for both CD-ROM and CD-WO. However, the cost of placing an archive on CD-ROMs is different from the costs associated with CD-WOs. The cost of producing CD-ROMs involves the cost of mastering, replication, and shipping the individual copies of a volume to users. The cost of making CD-WOs includes the cost of the CD writer system and the cost of blank CD-WO media. Figure 4 shows a comparison of costs for archiving data on CD-ROM and CD-WO media. Figure 4a is a plot of cost as a function of the number of copies for a single volume, whereas Figure 4b is a plot of cost as a function of the number of volumes. The graphs shows that for a small number of volumes, CD-ROM is more cost effective when more than a few dozen copies are needed. However, when the number of volumes is large (e.g., several hundred volumes), the cost savings of making only a few CD-WO copies instead of a CD-ROM master and many copies is quite significant.

There are some additional considerations relative to the use of CD-WO media and making only a small number of copies. Using CD-WO implies the availability of a central server to provide access to a copy of the data and a commitment to make copies or provide for electronic distribution to users. Other considerations of using CD-WO media are costs of network access and computer resources for the archive server and for the user. A central archive server would have the potential of being a single point failure. Also, heavy use of the server could lead to poor response of the system due to network limitations.

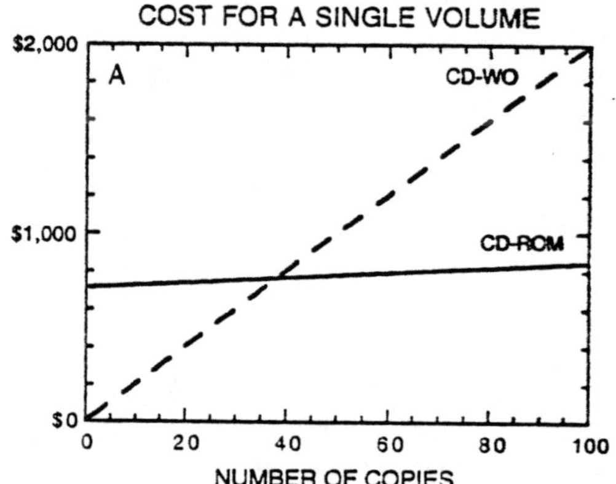

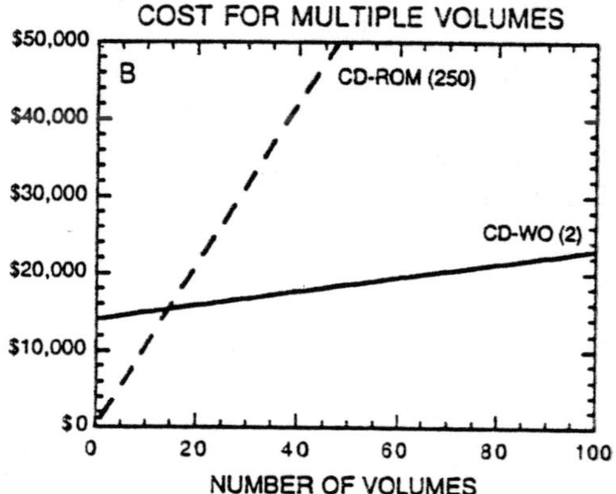

Figure 4. These two graphs compare the cost of archiving data on CD-ROM and CD-WO. (A) CD-ROM costs include the costs of mastering the volume and making copies, whereas CD-WO costs only include the media cost. (B) CD-ROM costs include mastering costs and the cost of making 250 copies of each volume, whereas CD-WO costs include the cost of a CD writer system and media costs for making 2 copies. Personnel costs associated with producing CD-WOs are not included in either graph. Dollar values are approximate since costs vary between vendors. However, the comparison is still valid.

Given the discussion above, we can establish some general guidelines for choosing between CD-ROM and CD-WO media for data archives. CD-ROMs are suited for datasets that are of a reasonable size, have data that are easily accessed, and have high use, which requires distribution of a large number of copies. Datasets that are very large, are complex and require special scientific or technical expertise for use, and are used by a small number of people are better suited for archiving on CD-WO media.

The guidelines listed above were used in determining how to archive the Magellan data. The F-BIDR dataset has a complex format and requires special software to access the data. The use of the F- BIDR dataset is relatively small, mainly for the construction of new F-MIDRs and extraction of detailed topography from stereo observations. Given these factors, coupled with the large size of the dataset (over 300 Gigabytes), it was decided to archive the F-BIDR dataset on CD-WOs. The collection of F-BIDRs requires over 500 CD-WO volumes. The data are distributed electronically, by making CD-WO copies, or by magnetic tape. Two

copies of each F-BIDR volume were initially made. One copy is used as a working copy at the PDS Geosciences Node to make copies when requested. The second copy is stored at the National Space Science Data Center (NSSDC) as a backup in case the working copy is ever damaged.

The MIDR dataset is stored in a simple image format. There is a large demand from the science community for the MIDR dataset, which is used in geologic mapping studies and regional to global scale investigations. The MIDR dataset is much smaller than the F-BIDR dataset. As a result the MIDRs are archived on a set of 126 CD-ROMs, with hundreds of copies distributed to scientists, libraries, educators, and the public. The altimetry, radiometry, and gravity datasets are also relatively small datasets and have widespread use in studies of Venus' surface and interior properties. These datasets are archived on about 22 CD-ROMs, and have also been widely distributed.

MAGELLAN STANDARD PRODUCTS CATALOG

Detailed information about the Magellan data products discussed above is maintained in the Magellan Standard Products Catalog. The standard Magellan products are those products that were planned, generated, and validated by or for the Magellan Project. The catalog component of the Magellan data access system is used to navigate through over 20,000 standard products in order to select a manageable subset of the data, based on criteria such as location, time, orbit, and proximity to a particular geographic feature. Once the desired products have been identified, the user may order them directly through the catalog. The catalog keeps track of where a product is stored (e.g., CD volume and file name), what latitude and longitude ranges it covers, when it was acquired, what source products were used to generate it, and other useful information. The catalog information is compiled from the ancillary information published on the CD-ROMs along with the data from the Magellan Project's internal product database.

The design of the catalog was based on requirements specified by a group of scientists and other users who represented the community interested in Magellan data. This group conducted formal reviews of the catalog after the design phase and after the implementation.

The catalog is organized as a set of relational database tables, shown in Figure 5. A table exists for each major entity for which information is stored, e.g., data products, features, orbits. Additional tables are created to represent relationships between entities. The tables are designed to minimize redundant information in order to save storage space and to avoid the

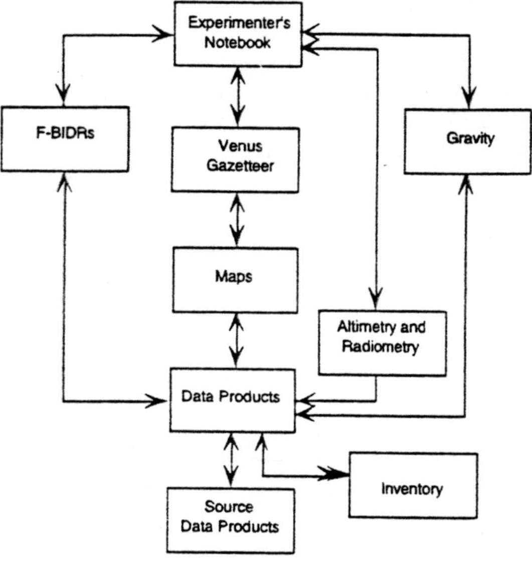

Figure 5. This diagram shows the conceptual organization of the Magellan Standard Products Catalog. Boxes are major entities for which information is stored. Lines are relationships between entities. Single-headed arrows show a one-to-one relationship. Double-headed arrows show a many-to-many relationship. The Inventory entity represents the different storage media for products.

maintenance difficulties involved when multiple copies of information are stored in different places. In the database design, Magellan products are grouped according to type — SAR image data, image maps, time-series data — in order to implement different search and information storage requirements for each product type.

The catalog is implemented using a commercial database management software package. The forms interface that is used to access the catalog data is implemented using tools provided with the database software. The interface can be used on a text terminal or an X-windows terminal. The user does not need to be aware of the underlying database structure in order to search the catalog. A user can choose among several methods for conducting a search (Figure 6). The simplest search involves entering the name of a product (or a partial name with a wildcard) and retrieving all the information known about that product. Another method is to enter latitude and longitude limits and retrieve all products that intersect the given area. The catalog includes a gazetteer of named geographic features on Venus, with approximate latitude and longitude limits, which can be used to locate products covering a selected feature.

The catalog also includes information from the

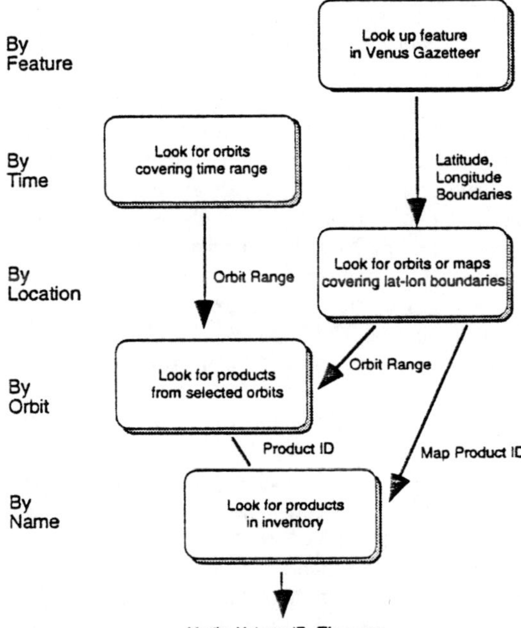

Figure 6. This schematic shows navigation through the Magellan Standard Product Catalog. A user can start with several search paths or categories, such as geographic feature, time, location, orbit, and product name. Connections between the categories allows the user to move through the catalog.

Magellan Experimenter's Notebook, a detailed record of the spacecraft state at regular intervals for each command upload. (A command upload is a set of instructions sent to the spacecraft). The Experimenter's Notebook can be searched by upload number, orbit number, or time, and resulting records can be used to locate data products.

For any search, the user may request a list of the products found, or simply a count of the products found, or he or she may browse through a multiple-page display of detailed information about each product found. For the image map products, the user may display a reduced-size contrast-enhanced version of the image in a separate window on the screen. Additional information available for most products includes the names of source products from which the products were generated, comments from the Magellan data processing team, names of features covered by the product, and for map products, details about the map projection used for each product.

A user may request delivery of products found in the catalog (Figure 7). Each time a desired product is encountered, the user adds it to a list of the products he or she wishes to order. The user can review and edit the list at any time during the catalog session. When satisfied with the list, the user enters his or her name, electronic mail address, and mailing address, and submits the order. The order is confirmed immediately by sending electronic mail back to the user. Orders that are delivered by post are normally processed by the next business day. If a user has requested electronic delivery, the order confirmation message will contain instructions for retrieving the data with ftp (a file transfer program available to many Internet users).

ELECTRONIC DATA DELIVERY

As discussed in a previous section, CD-ROM has proven to be an excellent medium for distributing the more popular Magellan products. CD-ROMs are not cost-effective for distribution of the very large F-BIDR dataset, which is of interest only to a small subset of the Magellan user community. Electronic delivery of a few products at a time, on demand, is a workable alternative for the F-BIDRs (Figure 8).

Automated electronic data delivery is facilitated by the use of CD-ROM jukeboxes. A 300- volume subset of the Magellan CD archive is in near on-line storage within the Magellan data access system using three CD-ROM jukeboxes. Each jukebox consists of a double speed CD drive flanked by two 50-slot disk packs. As needed, a jukebox is commanded to load a specified volume and the volume's contents are available to the system. If the requested volume is not present in one of the jukeboxes, an operator can easily replace a set of CDs in the jukeboxes. Currently, the 300 slots in the jukeboxes are filled with F-BIDR CD-WO volumes.

When a user places an order through the Magellan catalog, the information is added to an order database and a message is sent by electronic mail notifying the order manager of a new order (Figure 8). For items that are to be delivered by post, the order manager is responsible for seeing that the delivery is carried out, usually by the next business day. Electronic deliveries, however, are almost fully automated. A product marked for electronic delivery is located in the jukebox and copied to a staging area on hard disk, where it will remain for three days to give the user time to copy it. A mail message is automatically generated and sent to notify the user that the product is ready. The user can then retrieve the product at his or her convenience using anonymous ftp; the confirmation message that he or she received immediately after placing the order included instructions for using ftp (Figure 8). The order manager receives copies of all messages sent to users, and thus is informed of automatic deliveries. The order manager can intervene manually at any point in the process.

The successful automation of the electronic delivery system depends on the management of the three 100-slot jukeboxes. The jukeboxes are treated as a system resource such as a printer, and are managed by a queuing system. When the Magellan data access system receives an order for electronic delivery of a product that is on a CD in a jukebox, it makes an entry in the jukebox queue for that product. (Other programs besides the catalog may also make entries in the queue.) A jukebox manager program runs automatically at intervals throughout the day to process entries in the queue.

For each queue entry, the jukebox manager program performs the following steps. First, it checks that the requested volume is present in a jukebox. If the volume is not present, a message is sent to an operator with instructions for inserting the volume, and the queue entry is marked "pending," to be handled again the next time the queue is processed. If the volume is present, the program attempts to mount it, and if necessary will wait awhile until the jukebox becomes available for mounting. (After waiting a certain period of time, the program will give up, mark the entry "pending," and go on to the next entry.) When the volume has been mounted, the program copies the requested product to a reserved staging area on magnetic disk, and then dismounts the volume. The program then sends a message to the user who requested the product, notifying him or her that the product is ready to be copied with ftp. Finally, the program removes the entry from the queue, and goes on to the next entry.

In this system, the only time operator intervention is required is when a volume needs to be inserted in a jukebox. Since together the three jukeboxes hold 300 volumes, this does not happen frequently. The operator may intervene at any time, however, by editing the queue or starting the jukebox program manually. The operator receives copies of all messages that the jukebox program sends, so that he or she is always aware of any jukebox activity. The jukebox program records every action taken during the processing of the queue in a permanent log file, which the operator can use to check on the progress of an entry.

FUTURE DIRECTIONS

In summary, the experience of developing a system to access the Magellan data archives has led to the following conclusions. The most efficient way to distribute datasets that are in high demand is by archiving the data on CD-ROM and producing the appropriate number of copies. CD-ROMs are a high density storage media for relatively low cost, and they should be stable over many years. Datasets that are to be widely used from CD-ROM should be stored in formats that are relatively simple to ensure easy access to the data. Not all datasets, however, need to be archived on CD-ROM. Datasets that are very large, used less frequently, stored in complex formats, and that require specialized software in order to access the data are better suited for archiving on CD-WO. A small number of CD-WO copies (i.e., less than 10) can be more cost efficient than factory-produced CD-ROMs. However, archiving data on CD-WO implies the availability of a facility that can make copies of data for users. The approach used for Magellan data stored on CD-WO is to provide for electronic transfer of data from a set of CD jukeboxes. In any case, whether the data are stored on magnetic tape, CD-ROM, or CD-WO, users will need tools to search and select appropriate portions of datasets.

Over the next 5 to 10 years there will be several planetary missions that should produce archives of comparable size to the Magellan archives. For example, the completed Clementine Mission to the moon (Nozette, 1994) returned nearly 2 million separate images that will be archived on about 150 CD-ROM volumes. In addition, the Mars Surveyor Mission will be launched in 1996 with a camera system, a laser altimeter, and a thermal emission spectrometer. Each of these instruments will produce datasets of great interest to planetary geoscientists. The data archives from this mission will likely be similar in size to the Magellan data archive. Additionally, both the U.S. and Russia have plans to send several more spacecraft to Mars at intervals of every 2 years starting in 1998. The archiving philosophy used for Magellan and summarized in the previous paragraph will be used as guidelines for archiving data from these missions. The Magellan data access system is expected to serve as a model for data distribution systems for archives from these missions.

The current data access system used for Magellan will

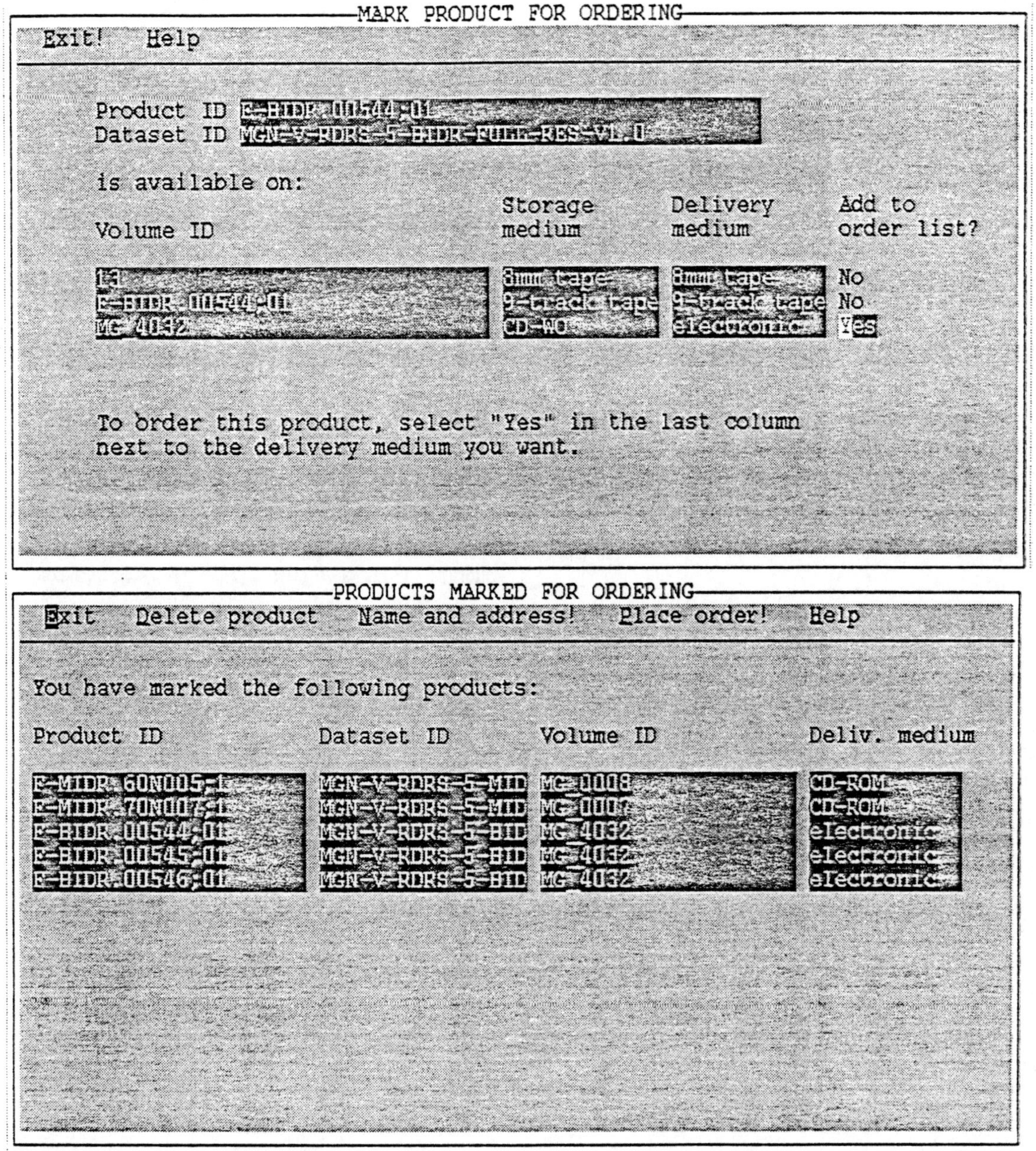

Figure 7. Examples of the Magellan Standards Product Catalog order screens. (Top) When a catalog user selects a product to order, this screen appears. In this example, F-BIDR.00544;01 (version 1 F-BIDR for orbit 544) is stored on three types of media: 8mm tape, 9-track tape, and CD-WO. The user has selected the CD-WO version to be delivered electronically. (Bottom) After marking products to order, the user can review the list on this screen. When satisfied with the list, the user chooses "Name and address" from the menu in order to enter his or her address. The user chooses "Place order" to submit the order.

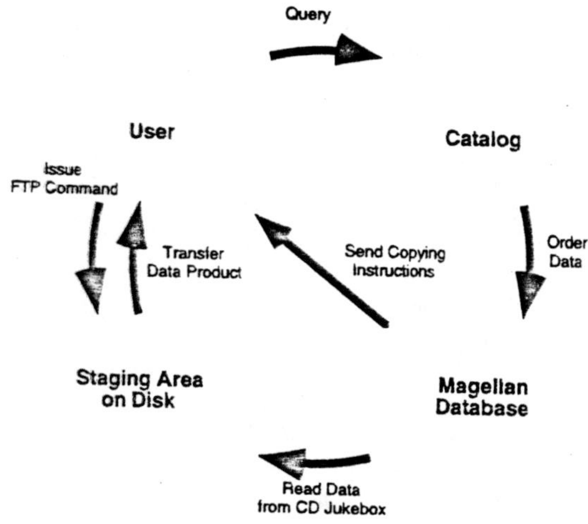

Figure 8. This diagram shows how data is ordered and received electronically through the Magellan data access system. The user starts with a query to the catalog to select a subset of products, with results submitted as an order. Data products are copied to a staging area, and the user receives an electronic message containing copying instructions. The user then retrieves the data.

have to continue to evolve to meet the needs of scientists as more and more missions return increasing amounts of data. Data navigation, the ability to find out what data are available and to select small, manageable portions of datasets, will become even more important. The development of map-based, graphical search tools would be a useful addition to the current methods of catalog searching. Finally, electronic distribution for large datasets is likely to become more common. Issues to consider for electronic distribution of large, on-line datasets are whether networks of the future will have the capacity to transfer increasing amounts of data and whether the networks can provide acceptable rates of response with increasing numbers of users.

ACKNOWLEDGMENTS

The authors are supported by JPL Contract 958756 to Washington University for PDS Geosciences Node tasks.

REFERENCES

Arvidson, R.E., and Dueck, S.L., 1994, The Planetary Data System: Remote Sensing Reviews, v. 9, p. 255-269.

Ford, P.G., and Pettengill, G.H., 1992, Venus topography and kilometer-scale slopes: Journal of Geophysical Research, v. 97, p. 13,103-13,114.

Nozette, S., 1994, The Clementine mission: An overview [abs.]: EOS, Transactions, American Geophysical Union, v. 75, p. 221-222.

Pettengill, G.H., Ford, P.G., and Wilt, R.J., 1992, Venus surface radiothermal emission as observed by Magellan: Journal of Geophysical Research, v. 97, p. 13,091-13,102.

Pettengill, G.H., and others, 1991, Magellan: Radar performance and data products: Science, v. 252, p. 260-265.

Saunders, R.S., and others, 1992, Magellan mission summary: Journal of Geophysical Research, v. 97, p. 13,067-13,090.

Saunders, R.S., and others, 1990, The Magellan Venus radar mapping mission: Journal of Geophysical Research, v. 95, p. 8339-8355.

Sjogren, W.L., and Konopliv, A.S., 1994, Venus gravity field determination: Progress and concern [abs.]: Lunar and Planetary Science XXV, p. 1281-1282.

DEVELOPMENT AND USE OF GIS APPLICATIONS IN THE GEOSCIENCES

Kym Pappathanasi-Fenton
William L. Hamilton

Department of Geography
Salem State College
Salem, Massachusetts 01970

Abstract – The use of Geographic Information Systems (GIS) technologies have increased in both number and format within specific geoscience specialties. The theme of this presentation is to detail, by way of a lecture with slides, the use of GIS within the domain of the geosciences. Investigating which geoscience sub-fields are the predominant users of GIS technologies, how GIS is being used, and what has been the rate of GIS integration throughout the geosciences in the past ten years, will be examined and presented. Answers to these questions will allow GIS professionals and geoscientists to identify opportunities for further research and development. Methodology included extensive manual and computer library searches developed as cascading categorization of GIS use and implementation. Initial results indicate that geoscience employ GIS for either descriptive geologic studies or explanations of how to establish a GIS for a geographic area but few are using it for advanced spatial analytic procedure.

INFORMATION SOURCES TO HELP EARTH SCIENCE STUDENTS LOCATE EMPLOYMENT OPPORTUNITIES

Julia H. Triplehorn

Librarian
Geophysical Institute
University of Alaska
Fairbanks, Alaska 99775

Abstract — Where do you find job advertisements in earth science? This paper will provide information on the key sources students should check as they begin their employment search. These will include journals which list job opportunities in earth science and related disciplines, professional association services, job hotlines, telephone book yellow pages, newspapers, books, and the Internet. The above sources will cover employment in industry, teaching at all levels, federal, state and international opportunities. This information will be of interest to job seekers as well as employers placing advertisements.

Job advertisements in earth science? Where do you find them? In libraries, these are usually asked questions not only by graduating seniors and graduate students who are seeking employment but also by faculty and personnel officers who want to advertise a position. To respond to these questions, the Washington State reference librarians a few years ago developed a clever solution by creating an in-house list of job advertisements found in the campus science journals.

This paper will provide an overview of job opportunities for earth scientists not only in science journals but will also additional sources. The following are the six areas to be covered:

1. Journals with job advertisements
2. Associations offering employment assistance
3. Job hotlines
4. Newspaper job advertisements
5. Telephone books
6. INTERNET employment services

JOURNALS WITH JOB ADVERTISEMENTS

Many geologists when asked about job advertisements in geology will suggest *Geotimes* or *EOS*. These are probably the primary sources. *Geotimes* job advertisements are frequently for earth science college or university teaching positions. *EOS* job advertisements include both teaching and research positions. These are divided into seven categories: atmospheric sciences, ocean sciences, geochemistry, solid earth geophysics, hydrology, space physics, and interdisciplinary. These two journals are not the only

ones which list job advertisements. My quest in this paper was to locate additional ones by checking *GEOREF's List of Priority Journals* and other sources.

Here are some of the results:

AAPG Explorer
Chronicle of Higher Education
Federal Career Opportunities
Federal Jobs Digest
GSA Today
Ground Water
Nature
New Scientist
Oil and Gas Journal
Photogrammetric Engineering and Remote Sensing
Science
Scientist

ASSOCIATIONS

Many of the journals which were examined in this project were issued by associations. The next question was the level of support in job searching provided by various geological associations to their members. Letters were sent to 30 geological associations found in the *Encyclopedia of Associations* and the *Geotimes* October Directory issue. Each professional society was asked if they offered any assistance to their members in locating jobs. Their responses are presented in Table I. (This table also provides additional current information, including telephone and Internet access).

Association	Internet	Journal or News-letter with Job Ads	Resume Assistance
AAPG American Association of Petroleum Geologists 918-584-2555		Paid classified	Employment interviews at annual meeting. Discount at Job Bank USA Career Advancement Service 1-800-2996-1USA
American Association for the Advancement of Science 202-326-7049		Employment Exchange Quarterly (free to members)	
American Geophysical Union 202-462-6900		EOS (weekly)	Job Center for registered meeting participants (spring and fall)
American Mining Congress 202-861-2800		AMC Journal	
American Women in Science 202-408-0742	awis@access.digex.net	AWIS Magazine	
Association for Women Geoscientists 612-426-3316	leete@macalstr.edu	GACA and local chapter newsletters	Assistance for members
Association of American Geographers 202-234-1450		AAG Monthly newsletter (primarily academic)	Convention Placement Service
Association of Engineering Geologists 508-443-4639		AEG News paid ads	Member resumes will be sent to section newsletters
Geological Society of America 303-447-2020	member@geo.society.org	GSA Today	
IEEE (Institute of Electrical and Electronic Engineers) 202-785-0017	info.ieeeusa.employ@ ieee.org		
North American Council on Geostatistics 303-273-3711		Geostatistics	
Northwest Mining Association 509-624-1158			Resume on file (informal) and resume binder at convention
SACNAS (Society for Advancement of Chicanos and Native Americans in Science) 408-459-4272	GOPHER:papilio.ucsc.edu TELNET:Papilio.ucsc.edu login:gopher		Recruiters at conference
Society for College Science Teachers			Employment Registry
Society of Economic Geologists 303-797-0332		List job opportunities newsletter	
South Texas Geological Society 512-822-9092		South Texas Geological Bulletin	Yes

JOB HOTLINES

In the library world, many regional associations and cities have regional job hotlines. This was one of the question I also included on the survey. There were no telephone hotlines found in the 30 associations. There are some government job hotlines which might have earth science related listings:

Department of Agriculture 301-344-2288
Department of Energy 202-586-4333
Geological Survey
(Eastern US) 703-648-7676
(Western US) 415-329-4122
Minerals Managment Service 703-787-1402

For a directory of additional joblines, try *Job Hotlines USA, The National Directory of Employer Joblines*, (Career Communications; 298 Main Street, P.O. Box 169, Harleyville, PA, 19438; $30.00). Your local public library may have a copy.

NEWSPAPERS

Newspapers are another traditional source for job advertisements. If the student has identified the area, where he or she may want to live, they can check the employment section of the area newspapers. Looking in *Ulrich's International Periodical Directory, volume 5, Newspapers* will help the student identify the area's major newspapers and their monthly subscription rates. Since it is arranged alphabetically by state and then by city, it is easy to use.

TELEPHONE BOOKS

Telephone books can be an additional source. Checking the yellow pages sections under geologists, consulting, mining, etc. may also give the student additional employment possibilities.

Phonedisc-Business CD, is a new national telephone yellow pages. It can be set to the industry desired and the state or city. This makes it easy to locate all the mining companies in Alaska or in Fairbanks and could be a very useful tool for students seeking work in particular areas.

Federal Yellow Book is a listing similar to the yellow pages which lists federal agencies and their employment division. If the student wants to apply at the national headquarters of the agency, as USGS-Reston, this yellow book will give the address and telephone number of the personnel office. Regional job listings can be obtained from the national headquarters. Some of the federal agencies (e.g., USGS) have job hotlines, as mentioned above.

INTERNET

Internet may have the answer to the reference question of where do you find job advertisements? Employment searching on the information highway is in its infancy as explained in a new book *Electronic Job Search Revolution*, by Joyce Lain Kennedy (New York, John Wiley and Sons, 1994, $12.95.) With 28 million worldwide Internet users, the net seems like a good place for listing employment opportunities. The following are a selected group of the currently available services on the Internet in late 1994.

Gopher to the Subject Oriented Internet Resource Guides (Univ. of Michigan); then locate the Employment Opportunities and Job Resources on the Internet. You will need a full Internet account (telenet and ftp, not just electronic mail) to access most of the services. Job listings are broken up into three groups:

1. Worldwide Newsgroups
2. Regional Hierarchy Newsgroups
3. International

The first has miscellaneous jobs; while the second lists jobs in particular regions of the U.S.; the third has international jobs. Many of these are free.

Here is a list of some available commercial services.

Academic Positions Network (Apple Tree Square Suite 1548, Bloomington, MN 55425; 612-225-1433) was started by the William C. Norris Institute in 1992 and lists faculty, staff, post-doctoral and fellowship positions. Positions are listed online for a one-time fee and positions are retained until requested to be taken off. The job searcher pays no fee to dial into the listing. Over 1000 institutions are members, so there are a broad spectrum of job opportunities. Access Academic Positions Network through these pathways: GOPHER: North America: USA: General: Academic Positions Network. TELNET: wcni.cis.umn.edu.

Career Connections (5150 El Camino Road Suite D33, Los Altos, CA 94022; 415-903-5815) is part of Heart Online Interactive Employment Network. Paperless resume management is their big selling point. Applicants can place their resume on the system and communicate totally, and even interview on the system. Employers pay to place advertisements on the system while the applicant pays no fee. Access HEART System through TELNET: college.career.com

E-Span - 3 years old (8440 Woodfield Crossing Suite 170, Indianapolis IN 46240; 800-682-2901, 317-469-4535) By accessing E-span through Compuserve, America Online, Genie, or Internet, you can browse job advertisements from Fortune 500 companies as well as other American, Canadian and international companies 24 hours a day. Listings are updated weekly. Candidates can submit resumes for a Job Search Resume Database which has both open and confidential registries.

Access E-Span Job Search through USENET: misc.jobs.offered. Note: Available newsgroups vary from host to host. Contact your system adminstrator if misc.jobs.offered is not available through your news server.

Federal Job Opportunity Board (Office of Personnel Management, Staffing Service Center, 4685 Log Cabin Drive, Macon, GA 31298) is an online bulletin board of current and expected job opportunities with the US government. These are listed by the job title and job series number and can be limited by state. It is updated every night which makes it very current. Access information is: OPM Job Bulletin Board, Modem: (912) 757-3100 (2400 baud,VT-100, parity: none, databits: 8, stop bits:1), TELNET: 198.78.46.10 or FLOB.MAIL.PM.GOV. Note: search limit options available include job title words, classification series numbers, and search by state.

CONCLUSION

This paper has been an overview to help students locate job advertisements in earth science and related disciplines. The key sources reviewed have been discipline journals with advertisements, associations offering employment assistance, job hotlines, newspaper advertisements, telephone book yellow pages and Internet resources.

PART III

POSTER SESSION:

GEOSCIENCE INFORMATION

SCIENTIFICALLY SIGNIFICANT GEOLOGIC FEATURES OF THE UPPER SNAKE ECOSYSTEM, SOUTHERN IDAHO

Terry Maley

Science and Planning Team
Bureau of Land Management
3380 Americana Terrace
Boise, Idaho 83706

Peter Oberlindacher

Geologist
Bureau of Land Management
3380 Americana Terrace
Boise, Idaho 83706

Abstract – In the Upper Snake Ecosystem, the Bureau of Land Management is conducting a pilot project to inventory and manage unique and scientifically significant geologic features and attractions. The results of this study will be used to develop similar projects in all the western states. The Upper Snake Ecosystem embraces an exceptionally large number of regionally and nationally significant resources characterized by outstanding examples of world-class tectonic features as well as a great variety of rock types, ages, and structures. The area represents a great natural outdoor geological museum and laboratory that is ideal for the study of the unique and diverse geology in one area.

INTRODUCTION

The Upper Snake Ecosystem, a drainage basin land management unit, embraces much of southern Idaho and small portions of Nevada, Utah and Wyoming. The Bureau of Land Management (BLM) has initiated a new program to inventory, describe and manage unique and scientifically significant geologic features and attractions (Maley and Randolph, 1993a; 1993b). Because the Upper Snake Ecosystem has an exceptionally large number and variety of these features including many that are nationally significant, this management unit was selected to showcase and implement the program nationwide. Nationally significant features include Menan Buttes, the Big Wood River, Bruneau Sand Dunes, Thousand Springs, Bonneville Flood features, Hagerman Fossil Beds, Silent City of Rocks, Shoshone Caves and the Great Rift and associated volcanic features.

First, we will establish draft standards and procedures to inventory, describe, interpret and evaluate these geologic attractions. Knowledge acquired from applying this process to a large variety of sites will enable us to later modify and complete the inventory and management process. This process will provide case examples of the basic types of attractions. These examples will be distributed throughout the bureau as models for nationwide consistency.

The management plan for each site will identify potential hazards and developmental potential for activities such as science education, interpretative centers, trails and pamphlets. The site reports will be abstracted to create a national listing for research scientists and educators. These sites will also be evaluated for their potential as Natural Landmarks, World Heritage Sites or sacred geography to American Indian religious beliefs.

OUTDOOR GEOLOGICAL MUSEUM

That portion of the Upper Snake Ecosystem situated in southern Idaho embraces a remarkably diverse group of scientifically significant geologic features. Furthermore, the regional geologic setting includes world-class tectonic features such as the basin and range faults, the Idaho-Wyoming thrust belt, the Snake River Plain and the Idaho Batholith. Rocks in the area range from 2200-year-old volcanic rocks at Craters of the Moon National Monument to the 2.7 billion-year-old gneiss in the Albion Range. The most recent geologic event to affect the area is the 15,000-year-old catastrophic Bonneville flood that swept across southern Idaho and left a variety of large-scale erosional and depositional features. Many of these unique geologic sites such as the Bruneau Sand Dunes, the Big Wood River, the Gooding City of Rocks, the Silent City of Rock, Shoshone Caves, Craters of the Moon and the lava fields along and associated with the Great Rift define ecosystems distinguished by a unique biota that developed as a direct consequence of the geologic environment. The unique occurrence of so many diverse scientifically significant geologic features within an area distinguished by a great variety of rock types, ages and structures has created a complete outdoor geological museum and laboratory. One of the benefits of this pilot project will be a directory and description of the geologic features, including maps, that can be used for field study by earth science students.

SCIENTIFICALLY SIGNIFICANT FEATURES

Thousand Springs and the Snake River Plain Aquifer

The Snake River Plain aquifer north of the Snake River is a remarkable aquifer of great resource and economic significance. The aquifer is not confined to a single homogeneous geologic formation. Rather it consists of basalt flows of the Quaternary Snake River Group. The individual flows are 3 to 5 m thick with the upper 1 to 2 m consisting of a very permeable rubble zone. Most of the rivers flowing from the mountain ranges some 65 to 75 km to the north of the Snake River, disappear into the surface of the northern edge of the Snake River Plain. The Little Lost River and the Big Lost River are examples of disappearing rivers. For about 160 km downstream from Milner Dam in the vicinity of Twin Falls an estimated volume of 5.8 km^3 of water enter the Snake River from gigantic springs on the north side of the canyon making it one of the world's most productive ground water systems (Whitehead and Covington, 1987; Malde, 1991). This is the well-known Thousand Springs area where the Snake River Canyon intercepts the Snake River Plain aquifer. The water table is general less than 150 m below the surface and the water-saturated basalt extends to a depth of about 1 km (Whitehead, 1986). According to Meinzer (1927) and Malde (1991) this 160 km reach of the canyon has 11 of the 65 springs in the United States with an average discharge exceeding 2.8 m^3/s.

Hagerman Fossil Beds

Abundant fossil remains of mammals and other vertebrates have been collected from deposits of the Glenns Ferry Formation. The best-known fossil assemblage has come from the "Hagerman Lake Beds" on the west side of the Snake River opposite the town of Hagerman. The fauna, mostly of horse-like animals, have been dated as Late Pliocene to Middle Pleistocene. The principal collections are from the "Hagerman horse quarry." The vertebrate remains found at the horse quarry include those of shrew, gopher, mole, weasel, otter, rabbit, peccary, camel, antelope, horse and mastodon, as well as those of fish, reptiles and birds.

Menan Buttes

The Menan Buttes are located on the Snake River Plain about 30 km north of Idaho Falls in southeastern Idaho (Hamilton and Myers, 1963; Maley, 1987a; Hackett and Morgan, 1988). The two cones, which consist of glassy olivine-basalt tuff, were erupted through the water-saturated flood plain of the Snake River. As a result of the buildup of the tuff during the eruption, the Henry's Fork of the Snake River was diverted to the south so that it now joins the Main Fork of the Snake River at the east edge of the south cone. The asymmetric shape of the cones was caused by prevailing wind which carried most of the pyroclastic fragments towards the northeast. Because the cones were erupted through the river bed, they contain many xenolithic cobbles, pebbles and sand grains of aplite, granite, quartzite, basalt and rhyolite from the river bed. The unique character of the tuff was caused by fluid lava making contact with the water in the river alluvium. Upon contact with water, the lava chilled to glass and was erupted explosively by steam.

Silent City of Rocks

The Silent City of Rocks covers a 17 km^2 area of the 28 million-year-old Almo Pluton in the southern part of the Albion Range in contact with an outlying gneissic dome complex that is more than 2.5 billion years old (Maley, 1987a). The spectacular rock forms are developed through a combination of jointing and weathering in the granitic rock. Joints control the basic shape of the landforms and facilitate the weathering process. Solutions migrate along the fractures and alter and disintegrate the surface layers of granite. Although jointing controls the general form of outcrops in the City of Rocks, spheroidal weathering with granular disintegration is the agent responsible for creating the bizarre and fantastic shapes that characterize the area. Case hardening of the outer surface of the granitic forms is also an important mechanism in creating unusual forms such as caves, niches, arches, bath tubs and sinks, hollow boulders and toadstools. The case hardening apparently occurred by iron oxides providing a protective crust to the surface. Once the hardened crust is penetrated, the much softer rock underneath can be quickly removed by weathering agents.

Gooding City of Rocks

The Gooding City of Rocks is situated about 28 km north of Gooding, Idaho. The landforms are composed of ash-flow tuffs about 10 to 11 m.y. old (Ekren and others, 1984; Dee, 1987a). Two sets of vertical joints intersecting at 90 degrees and differential weathering of the stratified tuffs create a forest of rock pillars or hoodoos 10 to 15 m high. Differential weathering of alternating hard and soft layers of tuff cause some pillars to look like stacks of coins. The most impressive forms occur in the drainage areas where water erosion has developed the tallest hoodoos.

Borah Peak Earthquake

On October 28, 1983, the Borah Peak Earthquake occurred in east-central Idaho on the west side of the Lost River Range, registering a Richter magnitude of 7.3 (Moore and Hoveland, 1987). The Borah Peak earth quake was the largest earthquake to hit the western United States in 24 years and was one of the only six historic earthquakes of magnitude 7.0 or greater in the Basin and Range recorded since 1872. The maximum vertical displacement on this normal fault was about 3 m on the west flank of Borah Peak. A minor component of left slip, indicates that the Lost River Range block also moved laterally northward, relative to the valley to the west. The surface deformation extends 37 km along the northern-trending range front and up to 100 m wide.

Catastrophic Bonneville Flood

The Bonneville Flood is one of the largest floods known from the geologic record. It is currently second to the Missoula or Bretz Flood which flooded the northwest more than 40 times. Pluvial Lake Bonneville once covered a vast area in northwest Utah to levels almost 335 m above the present level of the Great Salt Lake. Approximately 15,000 years ago, 4,700 km^3 of water (Malde, 1991) flowed over Red Rock Pass in southeast Idaho and then flowed westward, following the path of the present Snake River. The surface features of this flood, which were probably created in the first few days (Malde, 1968), are very similar in form and scale to those of the Missoula Flood. Some of the most impressive erosional features occur around Twin Falls and include marginal channels, scablands, coulees, dry falls (alcoves) and potholes. Depositional features left by the flood include huge deposits of Melon Gravels and large boulder trains composed of boulders more than 3 m in diameter. These melon gravel bars are up to 2.5

km long, 1.6-km wide and 100 m thick.

Bedrock Erosion in the Big Wood River

Potholes in the Big Wood River were first described by Russell (1902) and much later described in detail by Maley and Oberlindacher (1994). Approximately 10,000 years ago, near the end of the last glacial, the Big Wood River in the north-central Snake River Plain was rerouted by a large flow of basaltic lava from Black Butte. Since then the Big Wood River has developed remarkable bedrock erosional features in a 0.8-million-year-old basalt flow. These bedforms were sculpted by a large variety of igneous and metamorphic pebbles transported from the melting glaciers in the high country some 60 to 150 km to the north. The deeply incised inner channel of the Big Wood River reaches a depth of more than 10 m and a width of 4 m. The channel was developed by a series of aligned and coalesced potholes; and the morphology of the channel was created by potholes of various shapes and sizes and overprinted by asymmetric forms caused by the current.

The Great Rift and Associated Volcanic Features

The Great Rift system (Russell, 1902; Stearns, 1928; Murtaugh, 1961; Prinz, 1970; Kuntz and others, 1982, 1983, 1986, 1988, 1992) consists of a series of north-northwest-trending fractures, which extend 85 km from the northern margin of the eastern Snake River Plain to the Snake River. Shield volcanoes, cinder cones, lava flows and fissures are prominently exposed over a 2- to 15-km-wide belt along the Great Rift. Three lava fields are aligned along the rift: Craters of the Moon, Kings Bowl and Wapi.

Craters of the Moon lava field covers an area of 1600 km^2 and is the largest Holocene lava field in the United States. The flows were extruded during eight eruptive periods beginning about 15,000 years ago and ending about 2100 years ago. Craters of the Moon National Monument is a outdoor museum of volcanic features. Features such as crater wall fragments, cinder cones, mini-shield volcanoes, lava bombs, spatter cones, lava tubes, tree molds and rifts are among the best examples you can observe anywhere in North America. The King's Bowl and Wapi Fields, located at the southern end of the Great Rift system, were both erupted about 2,250 years ago.

Black Butte and the Shoshone Lava Flow

Black Butte is situated north of the town of Shoshone on the northern margin of the eastern Snake River Plain. It is a low shield volcano which rises approximately 70 m above the surrounding topography and has 3 km basal diameter. In the center of the shield is an irregularly shaped, 30-m-deep subsidence crater approximately 1 km in its largest diameter. The lava lake, which has an area of approximately 2 km^2, indicates a history of filling, overflowing and draining of degassed lava (Kuntz and other, 1992). Subsidence of what was once a lava lake occurred as the molten lavas flowed and drained through a 5-km-long, south-southeast-trending lava tube which is now largely collapsed. When the lake was at its highest, waves of molten lava washed over the crater rim depositing sheets of foamy pahoehoe on the flanks (Maley and Oberlindacher, 1993). The lava tube system contains both roofed and unroofed segments, leaving open trenches in places (Greeley, 1977). Many small lava tubes and channels distributed the lava away from the main tube (Greeley, 1982). After leaving Black Butte Crater, the flows followed the alluvial valleys of the Big Wood and Little Wood rivers to the south and then to the west, forming an L-shaped lava field some 60 km long. The 10,000-year-old Black Butte flow exhibits very little weathering, vegetation is sparse and the basaltic lava is still dark and crystalline in appearance (Maley and Oberlindacher, 1994).

The Big Lost River at Box Canyon

Box Canyon Gorge formed by late Pleistocene outburst floods from a glacial lake in the headwaters of the Big Lost River (Rathburn, 1993). Features caused by cataclysmic flooding include scour, cataracts, boulder bars and boulder trains. Cerling and others (1994) used cosmogenic ^{3}He and ^{21}Ne samples of the surface of basalt boulders to determine the flood occurred about 20,500 years ago.

Mackay Mining District

A textbook example of a small epizonal granitic pluton intruded through Paleozoic limestone is remarkably well exposed the Mackay Mining District. The gray limestone beds appear to be concordant and wrap around most of the granitic body (Maley, 1994). A well developed contact metamorphic deposit occurs at the limestone-granite contact (Umpleby, 1917) and skarn minerals such as quartz, epidote and garnet are plentiful. The area has many small mine dumps that yield a variety of collectable minerals.

Quartzite of Middle Mountain

In south-central Idaho a very unusual micaceous quartzite is mined from a group of quarries situated on the west flank of Middle Mountain (Maley, 1987a). The quartzite is unusual because it splits into large flat plates, more than 3 m in diameter along cleavage planes defined by biotite. The foliation or cleavage planes are consistently spaced at approximately 2 cm, yielding plates that do not vary in thickness over several meters.

The micaceous quartzite, sold under the trade name of "Rocky Mountain Quartzite" or "Oakley Stone," has been mined and sold in significant quantities since 1948. A national market was quickly established because it is much thinner than competing stone veneers; a ton of this quartzite will cover up to five times the surface area that can be covered by most competing veneer stones.

Bruneau Sand Dunes

The Bruneau dune field is located about 30 km south of Mountain Home, Idaho on the southern margin of the Snake River Plain (Baker and others, 1987; Murphy, 1973). Although there are many small dunes in the area, two large, overlapping sand dunes cover approximately 600 acres. These two imposing dunes are striking, particularly because they dwarf most of the nearby land features (Maley, 1987a). The westernmost dune is reported to be the largest single sand dune in North America, standing about 160 m above the level of the lake.

According to Murphy (1973) the largest dune is actually a dune complex or an aggregation of eolian features in a single structure. The stability of these light-gray dunes is attributed to wind blowing in two prevailing and opposing directions. Sand has probably collected in the basin since the basin was scoured clean 15,000 years ago by the Bonneville Flood.

Spencer Opal Deposit

The Spencer opal deposit is located about 8 km east-northeast of the town of Spencer. Many opal prospects lie on the south side of Opal Mountain; however, most of these deposits are covered by patented or unpatented mining claims. One of the best deposits of fire opal in the area is the well-known Deer Hunt Mine, reportedly discovered in 1948 by two deer hunters who were lost in a storm. People from all over the world visit this mine every year to collect the popular opal gemstone.

Opal occurs in cavities of a Tertiary-age rhyolite flow. The opaline nodules range in size from about 2 cm to 0.3 m. Opal occurs in the nodules as stratified cavity filling or closely spaced layers about 2 or 3 mm thick (Maley, 1987a).

Snake River Flood Gold

According to Wells (1969) the Snake River in southern Idaho has the best-known flood gold occurrences in the United States. Flood gold is finely divided gold that may be transported very long distances under flood conditions. The particles of such gold are so small that it may take 1,000 colors to be worth one cent, so as a general rule such deposits have little economic significance. Flood gold is distributed throughout the 1200-km length of the Snake River from the headwaters near Yellowstone National Park to Lewiston, Idaho (Dec, 1987b). Most workers have attributed the source of the gold to be late Mesozoic and early Cenozoic clastics in western Wyoming (Iddings and others, 1899; Antweiler and Lindsay, 1969; and Phillips, 1985). However, Desborough and others (1988) have found evidence for a local source of gold in the Blackfoot area. Gold occurs in the river bed, adjacent to and above the river. Unlike most gold placers where the density of the gold particles allows them to settle on or near bedrock, the Snake River flood gold is not concentrated on or close to bedrock.

Soldiers from Fort Boise first mined these Snake River deposits in about 1860 and a few years later some two thousand miners rushed to the upper Snake River to seek gold. The early miners referred to these deposits of gold as "skim bars" (Hill, 1916) because they tended to exist as thin layers at the surface.

Bruneau Canyon

The Bruneau River has eroded a spectacular narrow gorge commonly more than 300 m deep and with a rim-to-rim distance generally less than 1 km. The canyon starts about 15 km southeast of Bruneau, Idaho and runs almost due south more than 100 km across a volcanic plateau called the Bruneau-Jarbidge eruptive center (Bonnichsen, 1982). The walls of the canyon are steep to vertical and offer outstanding exposures of the volcanic flows. The volcanics of the Bruneau-Jarbidge eruptive center include the Idavada Volcanic rocks which consist of rhyolite ash-flow and lava flow units erupted during Miocene time. The silicic Idavada rocks are covered in places by more recent flows of Banbury Basalt also of Miocene time.

Bliss Landslide

On July 24, 1993, a landslide affecting an area about 100 acres occurred about 1 km south of the town of Bliss Idaho. According to eye-witness accounts the toe of the slide temporarily blocked the Snake River and forced the river to cut a new channel south of the slide. The movement over a 1-month period also wiped out almost 1 km of the paved Shoestring Road. The slide occurred in Pliocene sediments of the Glenns Ferry Formation which overlies pillow lavas of Tertiary Banbury Basalt. Glens Ferry sediments that make up the disrupted slide material are lacustrine clays of the Yahoo Clay (Malde, 1982). The Yahoo Clay was deposited when the McKinney Basalt dammed up the Snake River and formed a lake.

Outstanding textbook examples of most of the features typically associated with landslides can readily and easily be observed at the Bliss Landslide. These features include rotated slump blocks, hummocky surface at the earthflow portion near the toe, tension cracks, grabens, water ponds, staircase topography and a large head scarp.

Malad Gorge

Upstream from its confluence with the Snake River, the lower Big Wood River runs through Malad Gorge. This gorge is approximately 115 m deep and extends 3.5 km from the mouth of the Big Wood River. Where the Big Wood River makes its descent into Malad Gorge it is possible to observe current-driven pebbles swirling in potholes when the water level of the River is low. Malad Gorge was eroded when flows in the Big Wood River were much greater than at present. It was undoubtedly significantly enlarged by the action of the Bonneville Flood.

Saint Anthony Dune Field

The Saint Anthony sand dunes cover almost 300 km^2 in the area around Juniper Buttes near the town of Saint Anthony. The dunes were first described by Bradley (1873), were mapped in detail by Kuntz (1979) and described by Smith (1982). According to Baker and others (1987), most of the dunes are transverse and parabolic forms and are composed of sand derived from deposits of the Snake River and Henry's Fork. Older dunes stabilized by vegetation underlie the active dune field composed of migrating transverse and U-shaped dunes. Individual dunes reach a height of 100 m or more.

Island Park Caldera

Stearns and others (1938, 1939) first noted that the Island Park basin was a volcanic structure and Hamilton (1965) proposed that the basin is a caldera related to rhyolitic ash flows. Christiansen (1982, p. 345) found that "Island Park is the westernmost part of the Yellowstone Plateau volcanic field and is geologically transitional between the eastern Snake River Plain and the active part of the volcanic field." He also determined that Island Park is not a single caldera as suggested by Hamilton (1965). Instead, the Island Park basin has evolved over a 2-million-year period through three cycles of rhyolitic volcanism. Each cycle was characterized by an explosive eruption of thousands of cubic kilometers of rhyolitic magma. As each eruption left a partially empty magma chamber, the roof of the chamber collapsed and a caldera was formed.

REFERENCES

Antweiler, J.C. and Lindsey, D.A., 1969, Transport of gold particles along the Snake river, Wyoming and Idaho, *in* U.S. Geological Survey heavy metals program progress report 1968—topical studies: U.S. Geological Survey Circular 622, p. 6.

Baker, V.R., Greeley, R., Komar, P.D., Swanson, D.A., and Waitt, R.B., Jr., 1987, Columbia and Snake River Plains, *in* Graf, W.L., ed., Geomorphic systems of North America: Boulder, Colorado, Geological Society of America, Centennial Special Volume 2, p. 403-468.

Bonnichsen, B., 1982, The Bruneau-Jarbidge eruptive center, southwestern Idaho, *in* Bonnichsen, B. and Breckenridge, R.M., eds., Cenozoic geology of Idaho: Idaho Bureau of Mines and Geology Bulletin 26, p. 237-254.

Bradley, F.H., 1873, 6th annual report, U.S. Geological and Geographical Survey Territory: Washington, D.C., p. 211-212.

Cerling, T.E., Poreda, R.J., and Rathburn, S.L., 1994, Cosmogenic ^{3}He and ^{21}Ne age of the Big Lost River Flood, Snake River Plain, Idaho: Geology, v. 22, p. 227-230.

Christiansen, R.L., 1982, Late Cenozoic volcanism of the Island Park area, eastern Idaho, *in* Bonnichsen, B., and Breckenridge, R.M., eds., Cenozoic geology of Idaho: Idaho Bureau of Mines and Geology Bulletin 26, p. 345-368.

Dee, L.L., 1987a, The Gooding City of Rocks, *in* Maley, T.S., ed., Exploring Idaho geology: Boise, Idaho, Mineral Land Publications, p. 144-147.

Dee, L.L., 1987b, Snake River Gold, *in* Maley, T.S., ed., Exploring Idaho geology: Mineral Land Publications, Boise, Idaho, p. 148-150.

Desborough, G.A., Raymond, W.H., English, B.L., and Christian, R.P., 1988, Snake River gold in Idaho: distribution, grain size, grade, recovery and composition: U.S. Geological Survey Open-File Report 88-0352, 19 p.

Ekren, E.B., McIntyre, D.H., and Bennett, E.H., 1984, High-temperature, large-volume, lavalike ash-flow tuffs without calderas in southwestern Idaho: U.S. Geological Survey Professional Paper 1272, 76p.

Greeley, R., 1977, Basaltic "plains" volcanism, *in* Greeley, R., and King, J.S., eds., Volcanism of the eastern Snake River Plain, Idaho: A comparative planetary geology guidebook: Washington, D.C., National Aeronautics and Space Administration, p. 23-44.

Greeley, R., 1982, The style of basaltic volcanism in the eastern Snake River Plain, Idaho, *in* Bonnichsen, B., and Breckenridge, R.M., eds., Cenozoic geology of Idaho: Idaho Bureau of Mines and Geology Bulletin 26, p. 407-421.

Hackett, W.R., and Morgan, L.A., 1988, Explosive basaltic and rhyolitic volcanism of the eastern Snake River Plain, *in* Link, P.K., and Hackett, W.R., eds., Guidebook to the geology of central and southern Idaho: Idaho Geological Survey Bulletin 27, p. 283-301.

Hamilton, W., 1965, Geology and petrogenesis of the Island Park caldera of rhyolite and basalt, eastern Idaho: U.S. Geological Survey Professional Paper 504-C, 37 p.

Hamilton, W., and Myers, W.B., 1963, Menan Buttes, cones of glassy basalt tuff in the Snake River Plain, Idaho: U.S. Geological Survey Professional Paper 450E, p. E114-E118.

Hill, J.M., 1915, Notes on the fine gold of the Snake River, Idaho: U.S. Geological Survey Bulletin 620, p. 271-294.

Iddings, J.P., Weed, W.H., and Hague, A., 1899, Geology of the Yellowstone National Park: U.S. Geological Survey Monograph 32, part 2, p. 184-188.

Kuntz, M.A., 1979, Geologic map of the Juniper Buttes area, eastern Snake River Plain, Idaho: U.S. Geological Survey Miscellaneous Investigations Map I-1115, scale 1:48,000.

Kuntz, M.A., Champion, D.E., Spiker, E.C., Lefebvre, R.H., and McBroome, L.A., 1982, The Great Rift and the evolution of the Craters of the Moon lava field, Idaho *in* Bonnichsen, B. and Breckenridge, R.M., eds., Cenozoic geology of Idaho: Idaho Bureau of Mines and Geology Bulletin 26, p. 423-437.

Kuntz, M.A., Lefebvre, R.H., Champion, D.E., King, J.S., and Covington, H.R., 1983, Holocence basaltic volcanism along the Great Rift, central and eastern Snake River Plain, Idaho: Utah Geological and Mineral Survey Special Studies 61, Guidebook, pt. 3, p. 1-34.

Kuntz, M.A., Champion, D.E., Spiker, E.C., and Lefebvre, R.H., 1986, Contrasting magma types and steady-state, volume-predictable basaltic volcanism along the Great Rift, Idaho: Geological Society of America Bulletin, v. 97, p. 579-594.

Kuntz, M.A., Champion, D.E., Lefebvre, R.H. and Covington, H.R., 1988, Geologic map of the Craters of the Moon, Kings Bowl, and Wapi lava fields and the Great Rift volcanic zone, south-central Idaho: U.S. Geological Survey Miscellaneous Investigations Series Map I-1632, scale 1:100,000.

Kuntz, M.A., Covington, H.R., and Schorr, L.J., 1992, An overview of basaltic volcanism of the eastern Snake River Plain, Idaho, *in* Link, P.K., Kuntz, M.A., and Platt, L.B., eds., Regional geology of eastern Idaho and western Wyoming: Geological Society of America Memoir 179, p. 227-267.

Malde, H.E., 1968, The catastrophic late Pleistocene Bonneville Flood in the Snake River Plain, Idaho: U.S. Geological Survey Professional Paper 596, 52 p.

Malde, H.E., 1982, The Yahoo Clay, a lacustrine unit impounded by the McKinney Basalt of the Snake River canyon near Bliss, Idaho, *in* Bonnichsen, B. and Breckenridge, R.M., eds., Cenozoic geology of Idaho: Idaho Bureau of Mines and Geology Bulletin 26, p. 617-628.

Malde, H.E., 1991, Quaternary geology and structural history of the Snake River Plain, Idaho and Oregon, *in* Morrison, R.B., ed., Quaternary nonglacial geology: conterminous United States: Boulder, Colorado, Geological Society of America, The Geology of North America, v. K-2, p. 251-281.

Maley, T.S., 1987a, Exploring Idaho geology: Boise, Idaho, Mineral Land Publications, 232 p.

Maley, T.S., 1987b, Structural features in a veneer building stone deposit, south-central Idaho: Geological Society of America Abstracts with Programs, v. 19, no. 5, p. 318.

Maley, T.S. 1994, Field geology illustrated: Boise, Idaho, Mineral Land Publications, 316 p.

Maley, T.S. and Oberlindacher, P, 1993, Bedrock erosion in the lower Big Wood River channel, southcentral Idaho: Geological Society of America abstracts with programs, v. 25, no. 5, p. 113.

Maley, T.S., and Oberlindacher, P., 1994, Rocks and potholes of the Big Wood River, south-central Idaho: Idaho Geological Survey Information Circular 54, 41 p.

Maley, T.S. and Randolph, R., 1993a, New program to identify and manage unique and geologically significant resources on federal lands: Geological Society of America Abstracts with Programs, v. 25, no. 6, p. 313-314.

Maley, T.S. and Randolph, R., 1993b, Unique and geologically significant resources on federal lands: *in* Wick, C., ed., Proceedings of the Geoscience Information Society, v. 24, p. 197-204.

Meinzer, O.E., 1927, Large springs of the United States: U.S. Geological Survey Water-Supply Paper 557, 94 p.

Moore, S.W., and Hovland, R.D., 1987, Borah Peak Earthquake, *in* Maley, T.S., Exploring Idaho geology: Mineral Land Publications, p. 132-136.

Murphy, J.D., 1973, The Geology of Eagle Cove Basin at Bruneau, Idaho [M.S. Thesis]: Buffalo, New York, State University of New York, 77 p.

Murtaugh, J.G., 1961, Geology of Craters of the Moon National Monument, Idaho [M.S. thesis]: Moscow, Idaho, University of Idaho, 99 p.

Phillips, C.H., 1985, Intermountain gold anomaly; significance and potential: Engineering and Mining Journal, v. 186, p. 34-38.

Prinz, M., 1970, Idaho rift system, Snake River Plain, Idaho: Geological Society of America Bulletin, v. 81, p. 941-947.

Rathburn, S.L., 1993, Pleistocene cataclysmic flooding along the Big Lost River, east central Idaho: Geomorphology, v. 8, p. 305-319.

Russell, I.C., 1902, Geology and water resources of the Snake River Plains of Idaho: U.S. Geological Survey Bulletin 199, 192 p.

Smith, R.S.U., 1982, Sand dunes in the North American deserts, *in* Bender, G.L., ed., Reference handbook on the deserts of North America: Westport, Connecticut, Greenwood Press, p. 481-524.

Stearns, H.T., 1928, Craters of the Moon National Monument, Idaho: Idaho Bureau of Mines and Geology Bulletin 13, 57 p.

Stearns, H.T., Crandall, L., and Steward, W.G., 1938, Geology and groundwater resources of the Snake River Plain in southeastern Idaho: U.S. Geological Survey Water Supply Paper 774, 268 p.

Stearns, H.T., Bryan, L.L., and Crandall, L., 1939, Geology and water resources of the Mud Lake region, Idaho, including the Island Park area: U.S. Geological Survey Water-Supply Paper 818, 125 p.

Umpleby, J.B., 1917, Geology and ore deposits of the Mackay region, Idaho: U.S. Geological Survey Professional Paper 97, 129 p.

Wells, J.H., 1968, Placer examination; principles and practice: Bureau of Land Management Technical Bulletin 4, U.S. Department of the Interior, 209 p.

Whitehead, R.L., 1986, Geohydrologic framework of the Snake River Plain, Idaho and Oregon: U.S. Geological Survey Hydrologic Investigations Atlas HA-681, scale 1:1,000,000, 3 sheets.

Whitehead, R.L. and Covington, H.R., 1987, Thousand Springs area near Hagerman, Idaho *in* Buess, S.S., ed., Rocky Mountain Section of the Geological Society of America: Boulder, Colorado, Geological Society of America, Centennial Field Guide 2, p. 131-134.

INVENTORY OF LAND USE RESTRAINTS PROGRAM (ILURP): APPLICATION OF GEOGRAPHIC INFORMATION SYSTEMS IN MINERAL ASSESSMENTS

Neal B. Anderson
David A. Ferderer
Clark A. Roberts

U.S. Bureau of Mines, Intermountain Field Operations Center
P.O. Box 25086, Bldg. 20, Denver Federal Center
Denver, Colorado 80225

Abstract — In recent years, numerous laws, regulations, and management practices have been implemented to preserve or protect a variety of nonmineral resource values. As a consequence, Federal lands have been formally withdrawn from mineral exploration and development. These actions have led to concern that such restrictions may seriously impair access to our nation's mineral resources.

The U.S. Bureau of Mines is assessing the cumulative impact of federal land use decisions in 11 western states and Alaska. Each study is composed of three parts: (1) an inventory of Federal mineral ownership and mineral land availability, (2) an assessment of areas having past mineral production and/or known resources, and (3) a comparison of Federal mineral availability with areas having past mineral production and/or known mineral resources.

Geographic Information Systems technology is used to compile, analyze, and display the project results. In addition this technology lends itself to other implications, including regional issue identification and analysis, integrated resource management, binational border issues, resource versus reserve analysis, and multiple use conflict resolution.

USING A GIS TO MONITOR SEDIMENT RESOURCES IN THE COLORADO RIVER, GRAND CANYON, ARIZONA

Matt A. Kaplinski, David M. Best, Joseph E. Hazel, Jr.,
Jeffrey S. Wilkerson, Mark F. Manone, Alan R. Dale
Department of Geology, Box 4099
Northern Arizona University
Flagstaff, Arizona 86011-4099

Leland R. Dexter
Department of Geography and Public Planning, Box 15016
Northern Arizona University
Flagstaff, Arizona 86011

Sherry Jacobs, Hilary B. Mayes, Frank Protiva
Applied Technologies Associates, Box 22459
Flagstaff, Arizona 86002

Patrick J. Wright
Computer Data Systems, Inc., Denver Office
Denver, Colorado 80225

Abstract — A GIS is being developed to assist monitoring of Glen Canyon Dam impacts on the downstream ecosystem through the Grand Canyon. The objectives of the GIS are to provide a repository for historic and ongoing studies, provide a common platform for the integration of existing and future investigations, and to provide a database for long-term monitoring of changes in the Grand Canyon ecosystem.

A pilot GIS has been developed that extends from River Mile (RM) 60 to 72. This stretch includes the confluence of the Little Colorado River and the Colorado River at RM 61. The GIS database of the river corridor includes coverages of topography, survey control networks, gaging stations, surficial geology, vegetation, endangered species habitat, special study sites, and related tabular information.

Fluvial sediments deposited by the Colorado River are of particular interest because they form the foundation of the riparian ecosystem. For example, several study sites consist of topographic, bathymetric, and sedimentologic surveys which are repeated biannually. These surveys have been used at RM 62.4 to quantify the distribution and flux of sediments after natural flood events from the Little Colorado drainage. Hypsometric analyses and volumetric comparisons provide important field tests of flume experiments and numerically-derived models of sand bar evolution in eddy-dominated river systems. This pilot study demonstrates the utility of a GIS as both an analytical tool and a repository for the spatial-referenced information generated in these investigations.

INTRODUCTION

Glen Canyon Dam (GCD) operations have controlled the flow of the Colorado River through the Grand Canyon since its completion in 1963. Flows from GCD have severely impacted the downstream ecosystem.   The combination of reducing the downstream sediment supply by approximately 90% and releasing clear, cold water has threatened native plant and animal species, allowed exotic species to flourish, and eroded sandbars. In short, the effects of flow regulation from GCD have completely changed the downstream ecosystem. Public concern over these dam-induced impacts led the Department of the Interior to prepare an Environmental Impact Statement (EIS) on the effects of GCD on the downstream environment (GCD-EIS) and the Grand Canyon Protection Act of 1992. The Grand Canyon Protection Act requires the Secretary of the Interior to operate GCD "...under existing law in such a manner as to protect, mitigate adverse impacts to, and improve values for which Grand Canyon National Park and Glen Canyon National Recreational Area were established...". Development of a database for long term environmental monitoring is a critical component of future dam management strategies. Effective use of these data requires the development of an integrated and spatially correct database. The Bureau of Reclamation Glen Canyon Environmental Studies Geographic Information System (GCES-GIS) was created to address this requirement.

Sandbars are a primary natural and recreational resource of the river corridor because they form the foundation on which the fluvial ecosystem is structured. Therefore, sediment resources below GCD are a management priority. The purpose of this paper is to outline the structure of the GCES-GIS, describe the GCES-GIS site 5 pilot study, and present a GIS-based analytical application of the GCES-GIS at the River Mile (RM) 62 sand bar survey site.

THE GCES-GIS

The U.S. Bureau of Reclamation is the lead agency charged with preparing an Environmental Impact Statement on the impact of Glen Canyon Dam (GCD) operations on resources downstream in Glen and Grand Canyons. The Bureau of Reclamation Glen Canyon Environmental Studies office (GCES) in Flagstaff, Arizona, has taken the lead role in the development of a long-term monitoring plan that includes the development of a Geographic Information System (GIS) that will integrate the various aspects of research and monitoring in the Grand Canyon region (Werth et al., 1993). A GIS is an organized collection of computer hardware, software, and geographic data, designed to capture, store, update, manipulate, analyze, and display all forms of geographically referenced information (ESRI, 1993). The GCES-GIS will be focused on 17 study sites within the Grand Canyon (Figure 1). In order to measure changes over time in the environment, 0.5 meter topographic base maps and orthophotos of the river corridor have been produced by the Bureau of Reclamation Remote Sensing and Geographic Information Section for each of the 17 study sites. Environmental variables, such as vegetation and surficial geology, are mapped onto the topographic base and incorporated into the GCES-GIS.

Several studies are being conducted on sediment-related processes along the river corridor. Surficial geologic mapping of the GIS reaches is being conducted through Utah State University (Schmidt and others, 1994). Models of water flow and sediment transport are being constructed by the U.S. Geological Survey. Northern Arizona University, through the National Park Service, is monitoring the dynamics of individual sand bars at 30 sites located throughout the canyon (Cluer and Dexter, 1994; Kaplinski et al., 1994a). These investigations, as well as other historic information, will be incorporated into the GCES-GIS.

SITE 5 PILOT STUDY

A pilot study was initiated to determine the appropriate methodology prior to committing resources to all 17 study sites. Study site 5 extends from RM 60 to 72 and has been selected as the pilot study for the first phase of GCES-GIS development. Site 5 includes the confluence of the Colorado River and Little Colorado River (Figure 2). Site 5 was chosen as the pilot study because of the wide variety of investigations that are focused in the confluence region. As well as being the major source of sediment for the Colorado River system since Glen Canyon Dam closed its gates, the warm waters of the Little Colorado are a refugium for the endangered native fish populations, including the humpback chub *Gila Cipha*.

SAND BAR SURVEYS

The sandbar survey studies were designed to monitor the effects of Interim Flows (IF) on sand bars along the Colorado River through the Grand Canyon (Kaplinski et al, 1994a). The study of IF during the period of EIS review is important because the EIS Preferred Alternative (EIS-PA) closely resembles the IF (U.S. Bureau of Reclamation, 1994). The sand bar study involves the comparison of topographic and bathymetric surveys at 30 sandbars located in each of the 11 geomorphic reaches of the Colorado River corridor (as defined by Schmidt and Graf, 1990). In this example, a sandbar located within GCES-GIS site 5, at RM62.4 was chosen because it underwent significant changes during winter 1993 flood events that make it an ideal site to demonstrate the usefulness of the GIS as an analytical tool.

Flow Regimes

Interim flows have been in effect since August, 1991, and will continue until a Record of Decision is reached for the Glen Canyon Dam Environmental Impact Statement. Before IF, GCD was operated to yield maximum hydropower revenues through the production of peaking power. Flows released from GCD ranged between 85 m^3/s (3,000 ft^3/s) and 850 m^3/s (32,000 ft^3/s) daily depending on power demand. The IF limit the maximum discharge to 566 m^3/s (20,000 ft^3/s), the minimum to 142 m^3/s (5,000 ft^3/s), with upramp and

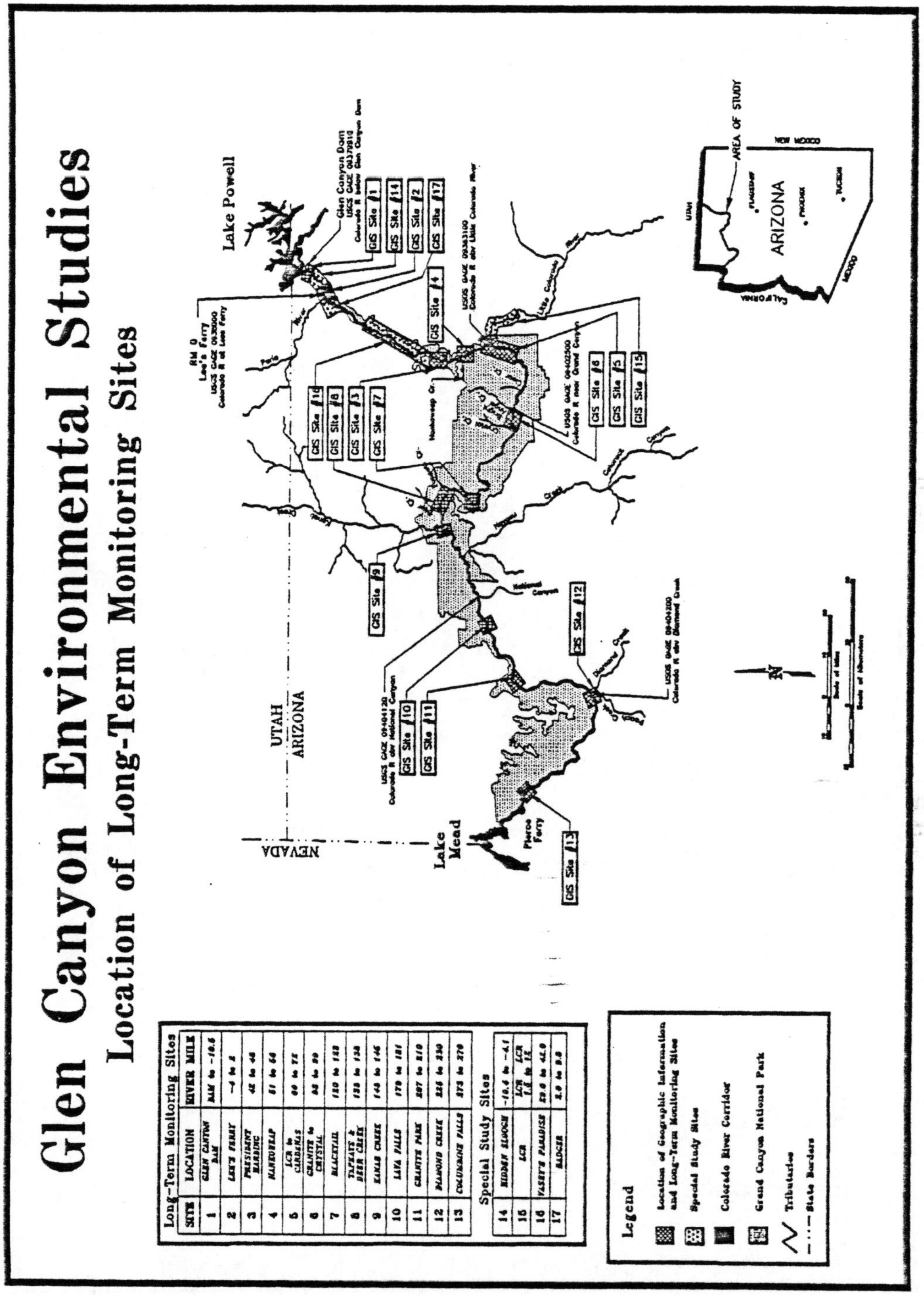

Glen Canyon Environmental Studies
Location of Long-Term Monitoring Sites

Long-Term Monitoring Sites		
SITE	LOCATION	RIVER MILE
1	GLEN CANYON DAM	DAM to -15.5
2	LEE'S FERRY	-4 to 5
3	PRESIDENT HARDING	42 to 45
4	NANKOWEAP	51 to 56
5	LCR to CARDENAS	60 to 75
6	GRANITE to CRYSTAL	65 to 90
7	BLACKTAIL	120 to 123
8	TAPEATS & DEER CREEK	133 to 138
9	KANAB CREEK	143 to 146
10	LAVA FALLS	179 to 181
11	GRANITE PARK	207 to 210
12	DIAMOND CREEK	225 to 230
13	COLUMBINE FALLS	273 to 276

Special Study Sites		
14	HIDDEN SLOUGH	-15.4 to -4.1
15	LCR	LCR 1.5 to LCR 15
16	VASEY'S PARADISE	29.0 to 42.0
17	BADGER	5.0 to 9.0

Legend

- Location of Geographic Information and Long-Term Monitoring Sites
- Special Study Sites
- Colorado River Corridor
- Grand Canyon National Park
- Tributaries
- State Borders

Figure 1. Location map of Glen Canyon Environmental Studies GIS monitoring sites.

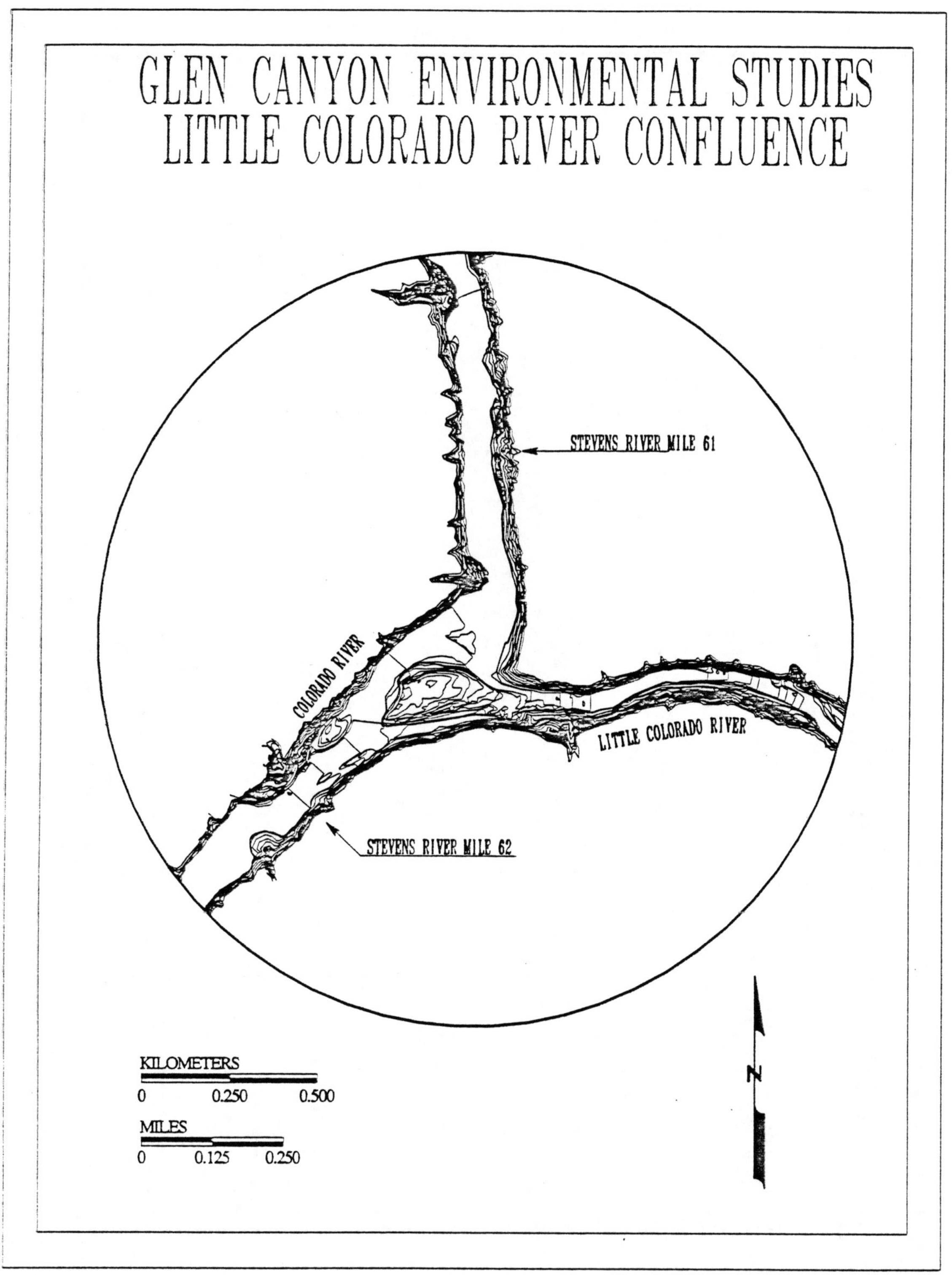

Figure 2. Topographic map of the Little Colorado River Confluence.

downramp rates of 57 m³/s/hr (2,000 ft³/s/hr) and 42.5 m³/s/hr (1,500 ft³/s/hr), respectively. Daily change cannot exceed 142 m³/s (5,000 ft³/s). These IF consist of low-, medium-, and high-volume months, with low flows during the late Spring and late fall, and high flows during mid-Summer and mid-Winter.

Natural flood events along the Little Colorado River during January and February, 1993, caused a significant deviation from the lower-volume IF regimes along the mainstem Colorado River (Figure 3). Three flood events occurred on the Little Colorado on January 12-16, January 19-23, and February 23-26, 1993, that delivered sediment and raised flows in the mainstem Colorado to 955 m³/s (33,777 ft³/s) , 783 m³/s (27,980 ft³/s), and 826 m³/s (29,450 ft³/s), respectively. Sand was deposited in nearly every eddy downstream of the Little Colorado-Colorado confluence for at least 30 miles, either adding to existing deposits or filling empty eddies.

RM62.4 Sandbar Study

A large sandbar located at RM62.4 was deposited mainly during the first and largest flood event (January 12-16, 1993) that provides a unique opportunity to examine the rate of sand bar development (Kaplinski et al., 1994b). Deposition at this site was consistent with published descriptions of depositional patterns at other Grand Canyon sandbars (Rubin and others, 1990,

Schmidt and others, 1994). However, before the floods there was no significant deposit present at this site. Therefore, this sandbar was deposited entirely during a single flood event, whereas most previously described sandbars are a composite of many different depositional packages. Detailed topographic and bathymetric surveys were conducted at this site in April, 1993, October, 1993, and April, 1994 . Comparison of the April 1993, and April 1994, surveys illustrates the pattern of post-flood erosion at the "Crash Canyon" site (Figure 4). In general, the greatest amounts of erosion (-4 to -11.5 meters) correspond with areas of maximum current velocity, particularly in the return current channel where currents sweeping across the eddy coalesce and are directed upstream.

We also observe that different bars have different susceptibilities to post-flood erosion and that post-flood dam operation exhibit some control on erosion rates at these sites. Approximately 5-10% of the "Crash Canyon" deposit we measured in April had been eroded by June, 1993. We estimate that erosion rates at the RM62.4 site ranged between 50 to 100 m³/day during this period of time. Dam operations changed from low volume (226 to 368 m³/s [8,000 to 13,000 ft³/s]) to high-volume (340 to 540 m³/s [12,000 to 19,000 ft³/s]) interim flows on July 1, 1993. Following this change in dam operations, the portion of the bar above the 142 m³/s (5,000 ft³/s) stage elevation had completely eroded within a two-to three-week period. Erosion rates during

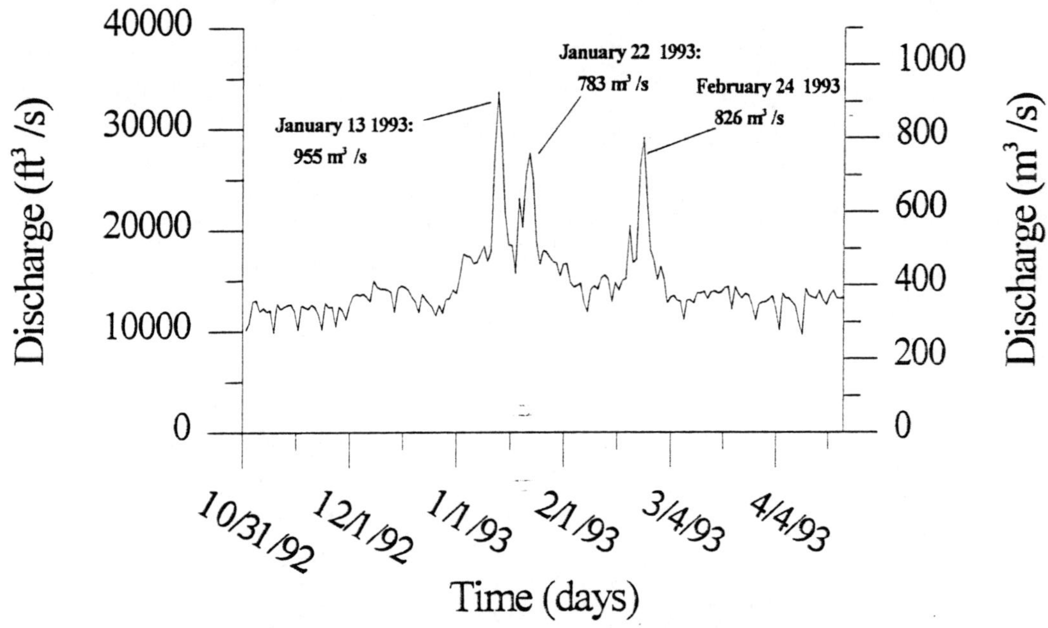

Figure 3. Colorado River hydrograph.

this period ranged from 2,000 to 2,500 m^3/day. Therefore, the change from low-volume to high-volume interim flow dam operations triggered an order of magnitude increase in erosion rates. However, because of the undercut, extremely unstable nature of the deposit, erosion rates may have increased regardless of changes in dam operations.

Hypsometric Analysis of the RM62.4 Sandbar

Hypsometric analysis relates the distribution of area and elevation within a given region (Strahler, 1952; Schumm, 1956; Bloom, 1991). By using dimensionless parameters, hypsometric curves can be used to compare different regions, irrespective of true scale. In this example we apply hypsometric analysis to an individual recirculation zone at the RM62.4 site. TIN models of topographic and bathymetric data are created for each survey and the areas enclosed by .5 meter contour lines are summed to construct hypsometric curves and histograms of area distribution (Figure 5). These data describe the change in sediment distribution within the eddy. This analysis will be repeated at all of the sand bar study sites and a statistical model relating the effects of area distribution on sedimentation patterns within recirculation zones will be constructed.

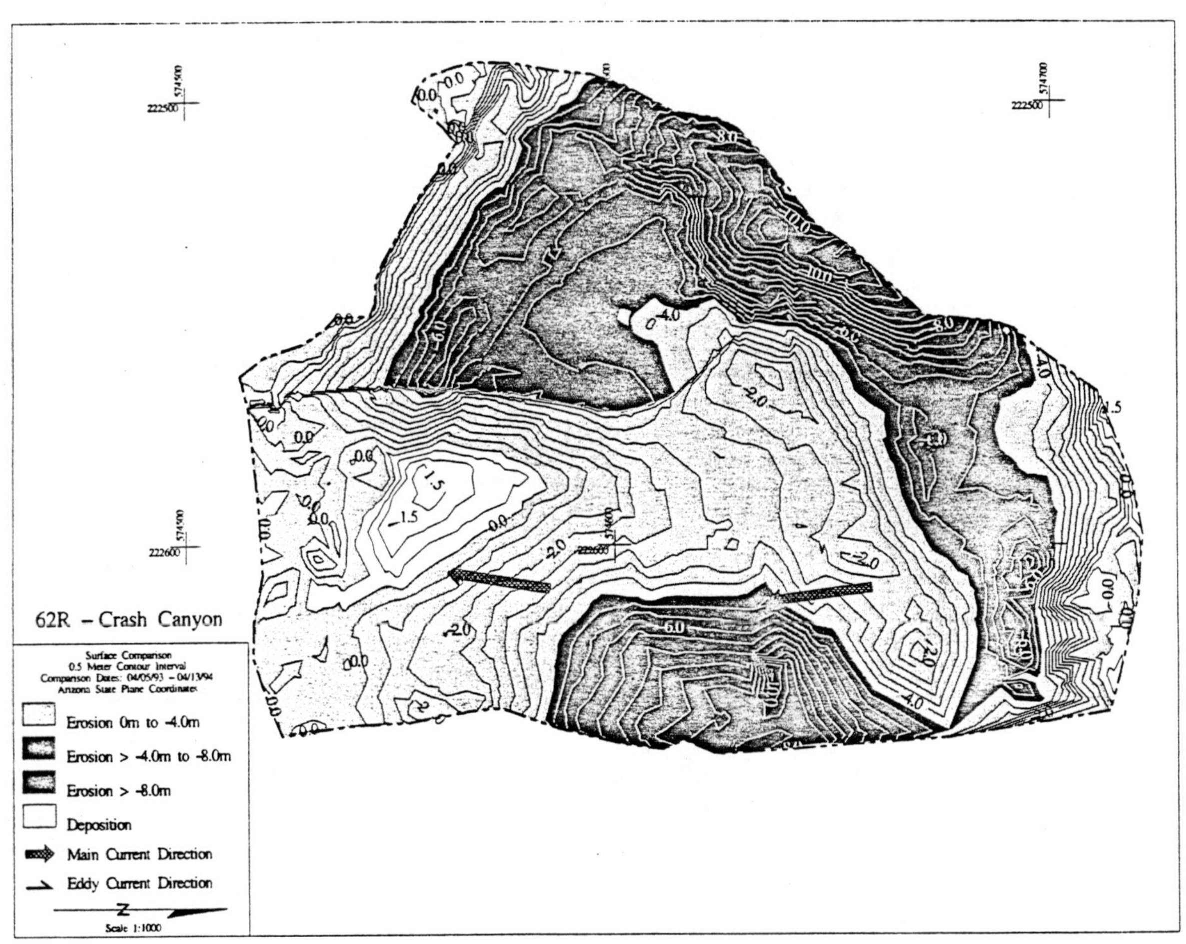

Figure 4. Contour map of the difference between the 4/5/93 and 4/13/94 topographic surveys at the RM 62.4 sandbar study site.

Comparison of recirculation zone area distribution describes the changing condition of fluvial sand deposits (Schmidt and others, 1993). The amount and distribution of sediment within an eddy has an important influence on the sediment flux between the recirculation zone and the main river channel (Beus and Avery, 1992). Describing the antecedent conditions of a recirculation zone through hypsometric analysis may prove to be an important predictive tool in determining where sediment may be deposited.

Discussion

Geographical Information Systems are an important tool for both scientists and environmental managers. Scientists involved in multidisciplinary projects need a centralized platform to archive, integrate, and analyze results from a wide variety of investigations. Scientifically-based land management strategies require that this information is in a well-documented and retreiveable form. GIS technology provides a solution to these problems.

The GCES-GIS was developed to assist the Bureau of Reclamation and Grand Canyon National Park in building an integrated ecosystem-based monitoring program. This system has integrated a wide range of investigations in the site 5 study area. GIS-based analysis of the RM62.4 sandbar study site has

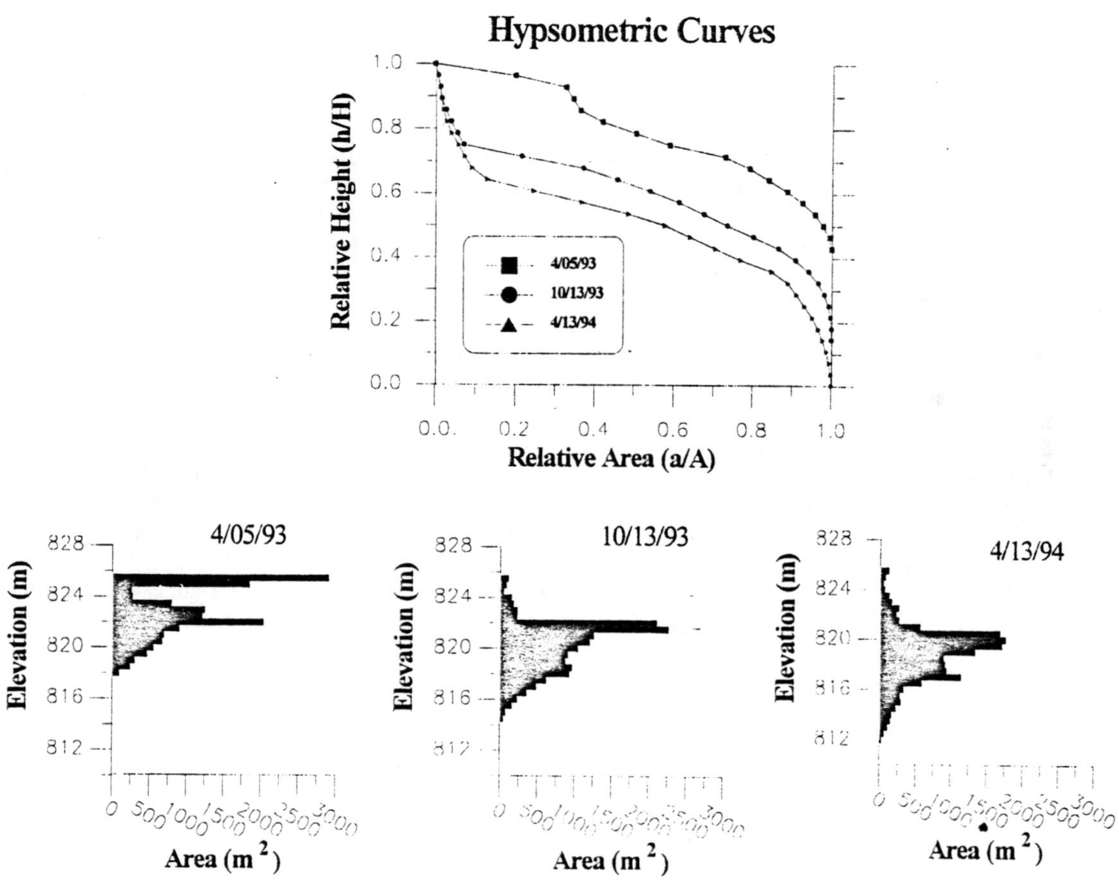

Figure 5. Hypsometric analysis of the RM 62.4 sand bar study site.

demonstrated the value of GIS systems in identifying sediment storage changes. The GCES-GIS should be an integral part of future monitoring plans for Grand Canyon National Park.

ACKNOWLEDGMENTS

Funding for this project was provided by the National Park Service (Cooperative Agreement # CA8034-8-0002) and Glen Canyon Environmental Studies office of the Bureau of Reclamation. Wendy Nelson provided invaluable secretarial and accounting assistance. Applied Technology Associates is a Government Services Administration (GSA) contractor to the Bureau of Reclamation Glen Canyon Environmental Studies office in Flagstaff, Arizona and Computer Data Systems Inc. is a GSA contractor to the Bureau of Reclamation Remote Sensing and Geographic Information Section in Denver, Colorado.

REFERENCES

Beus, S.S., and Avery, C.C., 1992, The influence of variable discharge regimes on Colorado River sand bars below Glen Canyon Dam: Final Report to the National Park Service, Cooperative Agreement # CA 8006-8-0002.

Bloom, A.L., 1991, Geomorphology: a systematic analysis of late Cenozoic landforms: Englewood Cliffs, New Jersey: Prentice Hall, 532 p.

Cluer, B., and Dexter, L.R., 1994, Daily dynamics of Grand Canyon sandbars: Monitoring with terrestrial photogrammetry: Final Report to the National Park Service, Cooperative Agreement # CA 8000-8-0002.

Environmental Systems Research Institute Inc., 1993, Understanding GIS - The ARC/INFO Method: Redlands, California, ESRI.

Kaplinski, M.A., Hazel, Jr., J.E., Beus, S.S., and Kearsley, L., 1994a, Monitoring the effects of interim flows from Glen Canyon Dam on sand bar dynamics and campsite size in the Colorado River Corridor, Grand Canyon National Park, Arizona: Draft Final Report, Cooperative Agreement # CA8022-8-0002.

Kaplinski, M.A., Hazel, Jr., J.E., Beus, S.S., Bjerrum, C.J., Rubin, C.M., and Stanley, R.G., 1994b, Structure and evolution of the "Dead Chub Eddy" sand bar, Colorado River, Grand Canyon: Geological Society of America Abstracts with Programs, v. 26, no. 6, p. 21.

Rubin, D.M., Schmidt, J.C., and Moore, J.N., 1990, Origin, structure, and evolution of a reattachment bar, Colorado River, Grand Canyon, Arizona: Journal of Sedimentary Petrology, v. 60, p. 982-991.

Schmidt, J.C., Mayes, J.L., Leschin, M.F., and Rubin, D.M., 1994, Geomorphic mapping of post-Glen Canyon fine-grained alluvial deposits of the Colorado River in rand Canyon, Arizona: Geological Society of America Abstracts with Programs, v. 26, n. 7, p. A-302

Schmidt, J.C., and Graf, J.B., 1990, Aggradation and degradation of alluvial sand deposits, 1965 to 1986, Colorado River, Grand Canyon National Park, Arizona: U.S. Geological Survey Professional Paper 1493, 74 p.

Schumm, S.A., 1956, Evolution of drainage systems and slopes in badlands at Perth Amboy, New Jersey: Geological Society of America Bulletin, v. 67, p. 597-646.

Strahler, A.N., 1952, Hypsometric (area-altitude) analysis of erosional topography: Geological Society of America Bulletin , v. 63, p. 1117-1142.

Werth, L.F., Wright, P.J., Pucherelli, M.J., Wegner, D.L., and Kimberling, D.N., 1993, Developing a Geographic Information System for resource monitoring on the Colorado River in the Grand Canyon: U.S. Department of the Interior, Bureau of Reclamation, Report # R-93-20, 46 p.

U.S. Bureau of Reclamation, 1994, Draft environmental impact statement, operation of Glen Canyon Dam, Colorado River storage project: Salt Lake City, Utah, 324 p.

PART IV

GIS PROFESSIONAL ISSUES FORUM

CLAYTONS LIBRARY:
GIS PROFESSIONAL ISSUES FORUM
GIS ANNUAL MEETING, SEATTLE, 1994

Margaret Eva

Geology Library
University of Queensland
Australia 4072

Abstract -- With the electronic library a reality, various questions need to be discussed, relating to costs, media, output, product quality, cooperation, clients' needs, hardware, and new ways of using information.

INTRODUCTION

This is a time to discuss professional issues which concern us. I feel very strongly that these issues are global, that we share these problems and interests, and that the best solutions will involve the wider community of geoscience information specialists.

There are a few issues where there is a unique Australian point of view; although these become fewer as there is less isolation every day. Two things come to mind -- we are very much at the mercy of currency variations, as we use the same publications and for us they are virtually all published overseas. We are concerned about critical scientific information being published outside the country, where we have no control over access or costs. We are also vulnerable to breaks in communication as greater commitment is given to electronic technology to communicate effectively across vast distances. Three times recently in Queensland the Internet connection has been severed -- once by a cut the cable in northern New South Wales, once by the failure of the Melbourne hub, and most recently by damage to the undersea cable off Tasmania.

Most of the following comments refer to academic libraries, and to geoscience information, even if there are wider implications.

CLAYTONS LIBRARY

Claytons library is the library you have when you don't have a library. (Claytons Tonic was a whiskey substitute, the drink you had when you were not having a drink). Patrons will sit in their own offices with a computer which provides access to library catalogues and which serves as the pipeline for actual information (articles, maps, interaction with colleagues or teachers), as well as the word processor and repository of their own research data. Since data can be transmitted by computer virtually instantaneously, it makes very little difference where the source and the end user are situated provided access to the information highway is unrestricted and free. The electronic journal or other electronically transferred information is not subject to postal delays, and will reach researchers all over the world at the same time, except for those who cannot afford the technology. The place of libraries in this process needs to be considered.

The America Physical Society has a vision of "integrating all the world's scientific literature into an electronic information system" (American Physical Society, 1991, p. 1133). There are many implications with more questions than answers. I don't want to sound negative because the future looks so exciting, but we must look hard at potential problems before making irrevocable decisions (or, by inaction, allowing them to be made for us).

I do not see electronic information as replacing the present contents of our libraries, at least not yet. We will need to cope with both, and I am not inclined to throw out what we now have.

The following are points which I suggest for discussion, based upon my ideas and imagination which do not necessarily reflect the official views of the Queensland University Libraries.

Costs

There is no reason to suppose that money will be saved in the short or long term by canceling current periodicals and redirecting the funds to document delivery for articles on demand (taking into account the costs of technical processing, storage, conservation and providing access to collections as well as the subscription).

Our experience with computer technology has been that it is always inadequate and obsolescent, replacement or upgrading always costs more than anticipated and (so far) has not actually saved effort, but has changed how we tackle projects. This automation has enabled us to provide services and information which would have been impossible with a manual system. The larger the files, the longer they take to search, and the more sophisticated the search software needs to be. There could be a significant cost just in regularly upgrading the access software, which would consume library budgets in the same way that traditional library materials, particularly serials, do now. How do we know that funds unavailable for the increased cost of subscriptions will appear for their digitized replacements? We do not yet know whether the cost (including overheads) of supplying only those articles which are requested, will be less than providing the whole title whether it is needed or not. That is not to say that it is not still worthwhile, if we can provide a better service.

Who will pay? The Australian Academic Research Network, AARNet, is to begin user charges next year. This is not likely to be on an individual basis, but I do not doubt we will be made to realise that access to the Internet is no longer free.

The cost of document supply from commercial vendors could follow the same path as serial prices — and we will have even less choice, if we have all cancelled our hard copies (supposing there still are printed journals). It reminds me of my concern over the leasing of CD-ROMs such as GeoRef — when we can no longer afford it, we will have nothing to show for all those dollars. Therefore I get anxious when I think of one or only a few sources of electronic journals, especially if they are commercial sites. I would rather think that universities could agree to share the load, perhaps on a subject basis, and also that we in Australia should not depend entirely on others. Even if we digitise our own products, they are unfortunately a very small proportion of our information needs. We also have to take extra costs of telecommunications into account, whether by fax or computer link.

Medium

I have great affection for books — they are portable, relatively durable, and you can own them. They require no extra equipment or power source, and they have the same words in them each time you go back to them.

Many people can read the same book, one after another, and the book is still there. Books are like people — all basically the same, yet no two are exactly alike. On the other hand, the words in the book are only accessible if someone actually reads them. If the text is digitised and added to a database, where every word is searchable and countable, and the information can be updated with a few keystrokes, all the contents are accessible, but not necessarily the same each time.

If one thinks about the CD-ROM as the alternative to online access, one soon finds that about 640 megabytes is very limiting. I have been told that the Ellis and Messina *Catalogue of Foraminifera* will require 15 to 17 CD-ROMs. This is a great improvement over the 11 shelves of cumbersome volumes, particularly if one has had to give up interfiling the supplements. The American Physical Society (American Physical Society, 1991, p. 1134) envisages 50 to 100 CD-ROMs for one year's output of physics literature (including "all published books, papers, conference abstracts and proceedings, numerical data, computer programs, etc."). Access even with a jukebox player will still be awkward.

Reading long articles on a computer terminal is very tiring and a strain on eyes. So there will be much printing out. A policy of not lending periodicals has the same effect. More printers and photocopiers will be needed, taking up the space released by the discarded printed books and periodicals.

Document Delivery

It is here now — Australian universities have free access to Current Contents and CARL Uncover for students and staff, who can order copies of articles online for faxing to their own office within twentyfour hours. We have fast track Interlibrary loans which promise the same. I was brought up on a cartoon of a "library" which consisted of a desk, a telephone, and a copy of the union catalogue. Now it is no joke, and the union catalogue is represented by a computer terminal. It is a sobering thought that it can be quicker to provide an article by fax from another library, than to produce an item which we own, but which is in storage. I have a vision of a mounting pile of articles, discarded because it is easier (though more expensive) to retrieve the article again than to find the original copy. Perhaps we have a role in teaching users how to deal with their mini libraries of faxed and downloaded information; we will have time, because we will no longer be managing the library collection.

There is a picture in my mind — Dr. Rockhead is interested in fossilite. He uses Mosaic to call up relevant files, locates twenty articles and orders them online with his (or his department's) credit card. Next day his fax extrudes about a hundred pages of low resolution black and white information. The half tone illustrations are not very clear, but he can still tell that 10 are just what he wanted, and he adds them to his home-designed filing system. The other 10 prove disappointing and he bins them, while in the next office his colleague Dr. Softhead requests them from the same source. Dr. Rockhead's bin is collected daily by the cleaner, who used to be the Geology Librarian before she was made redundant. Ms Helpyou is collecting all the discarded faxed and offprinted articles, and organising them (with an online catalogue). After six months the vendor of articles on demand trebles the prices and the university cuts off the credit card. Ms Helpyou offers to resell her articles for two-thirds of the commercial price and soon has a thriving business, which she calls a Library...

We need a way of telling which articles have been retrieved to the local site, so they can be shared. It will be possible to know exactly how often a particular article has been requested; I wonder how long those which are never or rarely requested will be retained, especially by commercial publishers.

We are already conscious that fax technology cannot yet cope with much more than text and basic line illustrations. Printing technology has only recently improved to the point where a substantial number of earth science publications regularly include colour illustrations, and maps have been a feature for over 100 years. Even half tone black and white illustrations cannot be adequately photocopied and therefore, faxed. There are already faculty at the University of Queensland who request a photocopy rather than a fax despite the extra delivery time.

International Transfer of Information

It is likely that most sites for electronic scientific information will be in the U.S. This poses no communications problems at the moment, except for the kind of hazards mentioned above, but could be a concern for countries such as Australia in the event of war or natural disaster. We should consider decentralized sites (as with Archie); this would also reduce the telecommunications traffic on the Internet.

Cooperation

Not only do we need to agree which collections will retain which titles, we should think of cooperative digitizing — i.e., agree among ourselves the priorities for titles to be retrospectively converted to digital form. We would base the decision on things like frequency of use,

shelfspace consumed, physical condition, suitability for conversion. Are we more interested in converting the little used but bulky titles (enabling patrons to continue playing with the originals of their favourite periodicals) or would the expense not be justified except for heavily used titles? A few years ago microfiche was thought to be the answer to space problems, but failed because of user resistance. We must be prepared for that again. On the other hand, when it comes to music we have happily progressed from 78 rpm to microgroove (and mono to stereo) to compact disc, from reel to reel tape to cassette tape to digital audio tape, without a pang. In Australia (particularly since 1988) we have been talking about the concept of a Distributed National Collection (Colloquium on the Distributed National Collection, 1993). We are using conspectus as a collection management tool, and are working on collection development policies.

Preservation and Completeness

With progressive digitization of serial titles, perhaps we should look again at binding and other conservation programmes. From the traffic in duplicates on GeoNet-L, it is clear that we all are still trying to complete holdings. If only one set is needed to serve the world, we should be able to save all that effort. We should also look at a programme to transfer complete runs of "technologically superseded" titles to less advantaged users. We must address the problem, instead, of preservation of electronic formats.

Undergraduates

It is easy to see universities organizing funding for academic (faculty) staff to use services like CARL Uncover and Current Contents, including hard copy or digital retrieval of data. It will change the whole process of funding and organizing undergraduate teaching if the same is to apply to first degree students who naturally outnumber other categories of academic users.

Access

Just as more users could access a card catalogue at the same time than can use a computer terminal, more patrons can use different volumes of a physical periodical run than can access the CD-ROM version. (Even if networked, there will be fewer terminals than physical volumes of the original).

I believe that a certain amount of information is retrieved by more or less accidental processes. We must ensure that we take advantage of the capacity of computers to organize and juxtapose concepts and ideas, and encourage new ways of browsing.

Attribution

When it is possible to run a computer search for a particular topic over the whole run of a periodical (or a number of them), how will we ensure that the bits of information retrieved will be (a) in context and (b) properly attributed to their author(s)? Also, some of the meaning is in the relationships between the documents, not just in their contents.

Repackaging and Filtering

Many geological insights come from a second or third look at original work (which may have been done for a different purpose in the first place). I have reservations about anyone in the information chain having the authority to decide for anyone else, what is important and what is not. Yet plainly, no-one can cope with everything in their own field of interest. For example, I go over my volumes of GIS proceedings every so often, and each time I find something which was not of great interest before, but is now just what I need.

With regard to the amount of new work being published, it seems to me that much of the cause lies with the academic system of publish or perish, and that university administrations should stop rewarding quantity over quality, and (in Australia, anyway) insisting on overseas publications as the only worthwhile vehicles for the results of research funded locally.

We are fortunately a committed and enthusiastic group, with formal links to the American Geological Institute; we are surely in a position to make our concerns known to the people who need us, and to act together (globally) to solve our problems.

ACKNOWLEDGMENTS

I wish to acknowledge the support of the University of Queensland Department of Earth Sciences and the University of Queensland Libraries, which made this visit possible.

REFERENCES

American Physical Society, 1991, Report of the APS Task Force on electronic information systems: American Physical Society Bulletin, v. 36, p. 1119-1151.

Colloquium on the Distributed Collection, 1993, Barr Smith Library, University of Adelaide: Australian Library Journal, v. 42, p. 3-12.